FORESIGHT AND KNOWLEDGE

FORESIGHT and KNOWLEDGE

Yves R. Simon

Edited by
Ralph Nelson *and* Anthony O. Simon

FORDHAM UNIVERSITY PRESS
New York
1996

Copyright © 1996 FORDHAM UNIVERSITY PRESS
All rights reserved.
LC 95–36922
ISBN 0–8232–1621–7 *(hardbound)*
ISBN 0–8232–1622–5 *(paperbound)*

Library of Congress Cataloguing-in-Publication Data

Simon, Yves René Marie, 1903–1961.
  [Prévoir et savoir. English]
  Foresight and knowledge / by Yves R. Simon ; edited by Ralph Nelson and Anthony O. Simon.
    p. cm.
  Includes bibliographical references and index.
  ISBN 0-8232-1621-7. —ISBN 0-8232-1622-5 (pbk.)
  1. Free will and determinism. 2. Knowledge, Theory of.
3. Science—Philosophy. I. Nelson, R. C. (Ralph Carl), 1927–
II. Simon, Anthony O. III. Title.
BJ1462.S4813 1995
194–dc20
                                                                95–36922
                                                                   CIP

Contents

Introduction	vii
Acknowledgments	xxi
Foreword	1
1. The Theory of Determinism	7
2. Science and Systematic Knowledge	67
3. The Vienna Circle	74
4. Epistemological Pluralism	90
5. Some Remarks on the Object of Physical Knowledge	100
6. Knowledge of the Soul	111
Bibliography	130
Index	137

Introduction

The fortunes of a book may depend on a number of factors other than its intrinsic merit. Among these factors are to be reckoned the language in which it was written, the publisher and its system of distribution, and the time it appeared. The French original of *Foresight and Knowledge* was published in 1944 by Éditions de l'Arbre, one of four new publishing houses that appeared in Quebec in 1941 (the others were Fides, Variétés, and Pony). The name (l'Arbre) "was a metaphor for the independence of French Canadian literature which no longer wished to be seen as a branch of French literature but as an autonomous tree."[1] Of course, it must be noted that these new publishers flourished at a time when France had been cut off from the Quebec literary market by the Occupation. Wallace Fowlie, the noted literary critic, wrote to Henry Miller recommending Éditions de l'Arbre in the following terms: "'There is a publishing house in Montreal of a liberal Catholic mind (in the tradition of Maritain) that publishes a number of good books.'"[2] He goes on to say that "'the two literary directors have waged a good fight against reactionary clerical forces.'"[3]

Not only did the Éditions de l'Arbre publish books, but the company also established a journal, *La Nouvelle Relève*, which lasted from 1941 to 1948. Yves R. Simon published a number of pieces in *La Nouvelle Relève* during the 1940s.[4] In the period following the Second World War Quebec literary publishing entered into a period of crisis brought on largely by the re-entry of the French in the Quebec literary market. As one observer put it, "'French publishers took back their rights.'"[5] While Éditions de l'Arbre published a list of new works in 1948, by 1949 none at all were listed.[6] *La Nouvelle Relève* as well ceased publication in 1948. Of the four debutantes of 1941, only Fides survived, and it continues to be an important publishing house today.

It seems that this brief account of the life and demise of the Éditions de l'Arbre is relevant to understanding the fate of *Prévoir et savoir*, often ignored when available and hardly available when it might have had some effect. The book was distributed in Canada and in the United States, but it would have been unlikely that it found its way to Europe during the war, and there is not much evidence that it was disseminated there afterward. It is likely, then, that it was denied an audience in France.

While parts of the book have appeared in English, it is only now, half a century later, that a translation of the book in its entirety is available. Had the original been better known, it is hard to believe that there would not have been considerable interest in an English translation. The author himself, so far as we know, did not initiate any translation, and at the time he made his own translation of the long first chapter, there were sections that he changed. He covered many of the issues examined in *Foresight and Knowledge* in a course of lectures given at the University of Chicago in 1959 entitled "Causality and Indetermination."[7]

The fact that the book was originally written in French does not explain the rather rare references to it by English-speaking philosophers, particularly Thomists. For any number of French texts are enumerated in bibliographies from which Simon's name is missing. Moreover, a perusal of a significant list of books about the philosophy of nature and the philosophy of science, especially works of a Thomistic bent or expressions of philosophical realism, works in which we might expect to find at least some mention of *Prévoir et savoir,* reveals that the title does not even figure in the bibliography. It can be said, I think, that it has been ignored by those who stand most to profit from it. Above all, the reader is astonished to find that *Prévoir et savoir* is totally ignored in tributes devoted to Yves Simon's philosophical *opera*. It is not even mentioned.[8]

Some might suppose that a treatise published fifty years ago is bound to be out of date, particularly in a domain where so many contributions have been made in the interim, the very influential writings of Karl Popper, for example, and of such New Philosophers of Science as Imre Lakatos, Thomas S. Kuhn, Paul Feyerabend, and Stephen Toulmin. Isn't a pre-Popperian treatise, to consider just one item, bound to be obsolete? The remarkable characteristic of *Foresight and Knowledge* is that it has avoided many of those features that tend to mark a book as a product of a certain time, place, and milieu. No doubt the amount of attention devoted to logical positivism, or logical empiricism, is one such mark, but most of the work displays an intellectual freshness, though a number of contemporary issues, such as the concern with inductive and deductive logic, are not addressed in it. It should be judged, rather, by the number of perennial issues, such as those concerning determinism, causality, and explanation, which are examined. And even though the attention once paid to logical positivism has subsequently been turned toward Popper, the reaction to logical positivism presupposes some acquaintance with it, as a dialectical movement requires knowledge of both thesis and antithesis. How otherwise can we understand Popper's falsification thesis than as a logical criticism of the verification thesis of the Vienna Circle? Simon begins by contrasting two conceptions of philosophy: (*a*) a type he refers to as literary philosophy and (*b*) his own conception of philosophy which stresses its kinship with science. In the chapter on Epistemological Pluralism, Simon, in dramatic contrast to the positivists, says:

> In our view, whoever wants to work out a theory of the relations between philosophy and the sciences should above all take note of the scientific character of philosophy and understand that metaphysics, which is the archetype of all philosophical thinking, is at the same time purely and simply the archetype of all scientific thinking [p. 91].

Somewhat earlier, in *The March to Liberation,* he referred to himself as "a man of science by profession."[9]

The author of a treatise in the philosophy of the sciences, if not an actual practitioner of scientific research, must deal with the frequent charge that philosophers are given to making judgments about scientific matters when they have little or no acquaintance with its practice, a notion that, if consistently pursued, would exclude all but practitioners from becoming involved in a metascientific discourse. In any case, Yves Simon was no armchair philosopher without direct acquaintance with the sciences, having devoted several years of his life to the study of them. We know that Simon not only studied the sciences, but, over and above his attachment to his philosophical vocation, even spent some time studying medicine, a fact that leads us to believe that the remark in the text about "feeling out of his depth" is autobiographical (p. 22). So we are not surprised when Simon says of the philosopher and the scientist, "we belong to the same breed" (p. 5).

In discussing the relations between philosophy and science, Simon stakes out a middle ground between two tendencies in contemporary philosophy. He supplies no examples of the literary philosophy he excoriates, but I suspect the phenomenon he had in mind is more prevalent now than when he made his reflections. A whole constellation of philosophers has become indistinguishable from sophisticated literary critics. I think particularly of the writings of Jacques Derrida in France and of Richard Rorty in the United States. For them there is no kinship between philosophy and science. If it is granted that science is about explanation, philosophy at best is interpretation, and then we have to discover what domain of interpretation is appropriate for philosophers, as opposed to biblical scholars, historians, and literary critics. If indeed philosophy is reduced to interpretation, yet has no specific domain, it is little wonder that there is much talk that philosophy has lost its identity, that it is coming to an end.

Now, there is a way of relating philosophy to science by asserting that philosophy comes out of, derives from, science. This is the way of Willard Van Orman Quine. "The philosopher's view is inevitably an extension of the scientist's. There is a continuity, if not an actual unity, of science and philosophy."[10] Simon does not believe that philosophy is derived from science any more than science is derived from philosophy. Even though he refers to the philosophy of scientists, "philosophy within positive science" (p. 48), meaning, I take it, the tendency of scientists to raise philosophical issues, he is clearly of the opinion that some scientists do it better than others, and, no doubt, some don't do it very well at all.[11] So he articulates a middle way between identification of the Quinean sort, philosophy is something scientists do, and the assertion of a total dissimilarity between philosophy and science. Philosophy and science are distinct; they are akin; they should not be separated. A simple way of stating the similarity would be to emphasize the common search for explanations.

Once that kinship is accepted, one is in a position to note the different ways in which the two proceed in this search. The differences are striking whether it concerns definition, determinism, causation, or concept resolution. Prior to the publication of *Prévoir et savoir,* Simon had distinguished between two paths to a knowledge of physical reality: an ontological path and another form following an empiriological or empiriometric path.[12] When naturalists define the human being using the cephalic index, its

upright posture, and the curvature of the vertical column (the spinal column), they do not furnish ontological definitions of the human, but an empiriological definition.[13]

As to determinism, Simon indicates that the term assumes different meanings depending on the system that is at stake. There are three different systems: that of common sense, that of science, and that of philosophy (p. 40). In like manner, if explanation means providing reasons or causes, we should not expect the same kind of explanation in philosophy and in science. On the one hand, there is the philosopher's quest for proper causes; on the other, the principle of causality undergoes a transformation, often expressed in terms of necessary and sufficient conditions (pp. 21–24). Indeed, in the positivist scheme of things, conditioning is completely substituted for the idea of causation. But to surrender the idea of causation is to give up the idea of explanation itself. Similarly, the attempt to convert the relation of causation into one of functionality in physical science would lead to the disappearance of the reality of physical change itself.[14] In conjunction with the analysis of causation, Simon notes as well the distinction between essential laws and statistical laws.

Finally, a distinction has to be made between the ways in which concepts are resolved in an ontological inquiry and in an empiriological one. Simon states the contrast very succinctly:

> Every positive scientist is on the wrong track when he uses a term whose meaning cannot be reduced to observable data. Every philosopher is on the wrong track who uses a term whose meaning cannot be reduced to being; the primary datum of the intellect [p. 76].

There are, then, a reduction to sensory evidence and a reduction to rational evidence.

A treatise in the philosophy of science (or the philosophy of the sciences) needs to be situated among the philosophical sciences. Several years before the publication of *Prévoir et savoir*, Simon gave its location as follows: "We believe that the part of philosophy which takes scientific knowledge as its object is only metaphysics itself in the exercise of its critical function. As to the philosophy of nature, its object is nature and not the science of nature."[15] In the original French version of *Foresight and Knowledge*, Simon has a phrase translated here as "a chapter of the critique of knowledge" (p. 39). In his own translation of Chapter One, he rendered *la critique* as "the philosophy of science"—an indication in my estimation that the two terms are synonymous. Because of the context, the present translation uses the term "the critique of knowledge" as appropriate.

Foresight and Knowledge is the result of the confluence of the two streams of Thomism and of French reflection on science from Auguste Comte to Émile Meyerson. In his extensive correspondence with Jacques Maritain, as yet unpublished, Yves Simon described *Prévoir et savoir* "as a seed in the furrow of *The Degrees of Knowledge*."[16] As commentators have noted, Simon is in the line of philosophical realism which, following Maritain's lead, he calls critical realism.[17] However, despite the emphasis on the critical approach as dominant, a noted phenomenologist, Aron Gurwitsch, insisted on referring to *Prévoir et savoir* as an example of naïve realism, a rather trite form of idealist put-down.[18] Gurwitsch believed that there was a linkage between phenomenology and the philosophy of science espoused by Ernst Cassirer.

The second stream consists of French reflection on science, the importance of investigations by Comte, Poincaré, Duhem, and Meyerson. Comte assigns as the goal of science not explanation, but mere prediction, human utility rather than the cognition of truth. Meyerson, on the contrary, maintained that formulating laws, general facts, was insufficient to constitute science, that science also attempts to explain phenomena.[19] Despite whatever reservations Simon had about Meyerson as a philosopher, and several times these reservations are voiced in *Foresight and Knowledge,* this emphasis on explanation accounts for Simon's consideration for him.

A notable omission in regard to French philosophy of science is the absence of any reference to Gaston Bachelard, noted critic of Meyerson,[20] although Simon had reviewed Bachelard's *L'expérience de l'espace dans la physique contemporaine* in 1937.[21] One might conjecture that Bachelard's Kantian-like form of idealism was repugnant to Simon or that perhaps Bachelard's approach offered little that was helpful to Simon's task. In any case the absence of any reference to him is remarkable. I would suggest that a useful comparison of the two philosophies of science, Bachelard's and Simon's, might focus on Simon's notion of *"the experimental absolute"* in "An Essay on Sensation,"[22] and the role of sensation, imagination, understanding, and construction in the constitution of scientific knowledge. What happens "if the experimental absolute is unqualifiedly negated?"[23]

Of course, it would be a distortion of Simon's explorations in the philosophy of science to see them as parochial, French parochial. It was the French perspective in which he was educated, but the reader will note the use he makes of Erwin Schrödinger and Max Planck, of Philipp Frank and Werner Heisenberg, to say nothing of Alfred North Whitehead and Sir Arthur Eddington.

An issue that has particularly preoccupied the French school after Comte is that of explanation and prediction in science. Meyerson's influence on Simon has already been alluded to. Perhaps more than any other chapter in the treatise, "Science and Systematic Knowledge" specifically addresses the connection between foresight and knowledge. Though brief, it indicates the direction a more complete treatment might fruitfully pursue. Comte, it is argued, had confused science and systematic knowledge. Corresponding to each of these kinds of knowledge is a kind of prediction. However, while "the prediction of phenomena is one of the essential functions of science" (p. 67), "[c]oncrete prediction is outside the realm of pure science" (p. 68), though it is what systematic knowledge is about, searching as it does for a practical solution. This is tantamount to saying that the pertinent contrast is between theoretical prediction and prediction related to practice. Stephen Toulmin in *Foresight and Understanding* makes a similar point when he says that "science is certainly not a matter of forecasting alone, since we have to discover also explanatory connections between the happenings we predict."[24] He rejects what he calls "the predictivist account of explanation"[25] and shows instances where explanations may not involve forecasting at all. Toulmin then emphasizes the disjunction of explanation and prediction, sometimes identified, while Simon wants to distinguish two different notions of knowledge to which characteristic forms of prediction correspond.

Turning from problems that arise in Comtean positivism to an examination of logical positivism, more precisely the Vienna Circle, Simon discusses its doctrines at a moment of its growing ascendancy in North America, which turned out to be an important phase in the history of the philosophy of science rather than a lasting school.

Although there were aspects of the new empiricism (or logical positivism) which Simon was bound to view negatively—notably the characterization of metaphysics—his approach to the Vienna Circle is positive, albeit critical, in that he seeks to learn what contribution its representatives have made to our understanding of scientific knowledge. In this regard, he devotes more attention to the lesser-known Hans Hahn than to the well-known Rudolf Carnap.

Concentrating on the new empiricism as a theory of meaning, Simon discusses semiotics and sign functions. He reflects on the process of concept resolution, the empiricist postulate and the exclusion of metaphysics, the relationship between observation and theory, and the pragmatic conception of truth endorsed by the new empiricism.

Insofar as logical positivism maintains that only positive science can furnish true knowledge, it is a brand of monism or, some might prefer, a brand of reductionism. It is obvious at this stage that Simon adheres to an epistemological pluralism. I would relate his comments in the fourth chapter to two discernible tendencies in modern Thomism—which is not to imply that there are not others—concerning the philosophy of nature. On the one hand, there is a school of metaphysics in which the philosophy of nature is completely ignored. The issue of its status vis-à-vis metaphysics is never discussed. On the other, there is another school or tendency which indeed asserts the importance of the philosophy of nature, but refuses to accept the notion of an autonomous science of nature. So either metaphysics absorbs the philosophy of nature or the philosophy of nature absorbs any other approach to the study of nature. It is apparent that Simon has a place for both an ontological and an empiriological knowledge of nature. And metaphysics again is a distinct form of knowledge.

Psychology, in Simon's view, is "a part of the science of nature."[26] It is not a social science. Since nowhere in *Foresight and Knowledge* is there an examination of the social sciences as such, it would be useful to look at some of Simon's reflections in an essay first published in 1953 and recently republished in *Practical Knowledge*.[27] The essay entitled "From the Science of Nature to the Science of Society" attempts to show why the natural science model is not an appropriate one for social science. At the first stage, when the notion of social science was emerging, the project was to produce "a science of society patterned after the science of nature."[28] Were such a science developed, its applied side would be social engineering, just as applied natural science is engineering. Then, at a later stage, Max Weber introduced the notion of value neutrality. Weber's argument entails the independence of social science from morality. The riposte to this conception, trenchantly stated by Leo Strauss, is that moral considerations enter into the very context of social inquiry. They are essential to understanding social, that is human, facts. Simon's own response to the Weberian notion of ethical neutrality is to counter Weber's distinctions with one of his own, that between nature and use. For instance, because psychology is a natural science, meaning that it is concerned with nature, not use, "The principle of ethical neutrality holds in psychology."[29] But it is quite different when considerations of nature and use are inseparable. Simon concludes that "Facts pertaining to the life of human society seem to be of such character that a philosophy of man is necessarily at work in the reading of their intelligibility."[30] This seems clearly to be a rejection of an autonomous social science, presumably a theoretical discipline. Consequently, the relation of social philosophy to social science is in no way comparable to the relation between the philosophy of nature and the science of nature. It might be comparable were one to conceive of social philosophy and social science as both theoretical inquiries; Simon denies that either of them is theoretical. Together they fall under the heading of practical knowledge.

The last two chapters contain companion pieces in the knowledge of nature: the first concentrates on the various definitions of physical knowledge; the second, on the epistemological confusion in contemporary psychology. The former treats the issue of what we know through perception and what we conceptualize. It is a corrective of a certain conception of perception, such as the Berkeleyan, with its failure to distinguish between *per se* and *per accidens* perception. Because of its brevity, the fifth chapter should be read in conjunction with Simon's "An Essay on Sensation."

In the recently published *The Definition of Moral Virtue,* based on lectures that Simon gave at the University of Chicago in 1957, there is a passage which helps us understand the place that the chapter on knowledge of the soul has in Simon's work as a philosopher.

Let us admit that psychology is a very poorly organized discipline and one whose disorderliness does not seem to be diminishing. Were I a little younger, I would consider dedicating my life to improving the situation, because the science of the soul is so important for morality. But sometimes I wonder if it is not already too late.[31]

This essay on the epistemological nature of psychology, which appeared in *Gants du Ciel,* No. 5, 1944, provides an introduction to that project envisaged by Simon but never completed. It is a starting point. It is not a constitutive epistemology in the Kantian sense but a reflective one, that is, it surveys the state of the discipline in order to clarify the problems, eliminate false leads, identify the areas requiring further development, and, finally, point out the interconnections. Simon's essay is timely in the absence of an undisputed and unified science of psychology commanding the support of competent persons. It is not dated, because he sticks to the issues and does not dwell on personalities. There are no references to twentieth-century psychologists in the text, but it is easy to supply topical references to illustrate a point, whether it be Sigmund Freud, Jean Piaget, or B. F. Skinner. Take, for instance, the attack on philosophical psychology. In Freud's case, there was a constant effort to completely divorce psychology from philosophy, which he distrusted, even though some would say that there is an implicit or disguised Freudian philosophy. In Piaget's case, there is a vehement polemic against philosophical psychology in his book *Sagesse et illusions de la philosophie* (1965).[32] It seems clear that Piaget would not acknowledge any positive contribution to psychology in the writings of Maine de Biran, Bergson, Sartre, or Merleau-Ponty, to mention just the better known. Philosophy might be "wisdom" in Piaget's sense, but that is far from being a compliment since "wisdom" is not "science." In the case of B. F. Skinner and his efforts to go beyond freedom and dignity, there is a perfect exemplification of that technology extended to human beings so aptly described by Simon. Yet in spite of scientistic confidence, the identity crisis in psychology persists.

If one is neither to reject philosophical psychology out of hand nor to dismiss the claims of positive psychology, it is necessary to make appropriate distinctions, such as the distinction between theoretical and practical psychology and between applied psychology and moral psychology. Yet, having done that, Simon shows that an analysis of individual cases reveals that applied and moral psychology are complementary.

Simon's contribution to the task of "improving the situation" did not end with the epistemological essay. It is important to note his treatment of moral psychology. I recall years ago reading St. Thomas's examination of human acts in the *Summa Theologica* and thinking that this was a kind of moral psychology. However, that was not then, nor is it now, a common term. In Rawls's *A Theory of Justice*, there is a section on moral psychology which seems to be about moral learning. A recent study by an English philosopher, N. J. H. Dent, *The Moral Psychology of the Virtues*,[33] seems more to the point; following G. E. M. Anscombe and analytic philosophy, it examines psychological concepts particularly relevant to ethics. Now, Simon had not only talked about moral psychology but also developed a number of concepts that are characteristic of it, such as the notion of practical reasoning, free choice, knowledge by inclination, and virtue (based on the important distinction between habit and *habitus*). Again, *The Definition of Moral Virtue* brings out the significance of Simon's contribution from the viewpoint of both method and content.

Much has been done, through the efforts of scholars like Vukan Kuic and others, to publicize the political philosophy of Yves R. Simon. Much remains to be done in developing a moral psychology with a Thomistic inspiration. In the project of developing such a systematized approach, Simon's writings could well furnish the basis and starting-point. Let us hope that the growing interest in and availability of Simon's philosophical works will encourage younger scholars to take up the task Simon did so much to promote.

RALPH NELSON

NOTES

1. *L'édition littéraire au Québec de 1940 à 1960,* ed. Richard Giguère and Jacques Michon (Sherbrooke, Québec: Université de Sherbrooke, 1985), p. 8.
2. Quoted in ibid., p. 11.
3. Quoted in ibid., p. 11.

4. Four of the articles appearing in February, March, April, and October were commentaries on the course of the war. A fifth was occasioned by Jacques Maritain's sixtieth birthday. In addition they published two segments of Simon's memoirs.

5. Quoted in *L'édition littéraire au Québec*, p. 16.

6. The information is gathered from the *Bulletin bibliographique de la Société des Ecrivains Canadiens* (Montreal: Société des Écrivains Canadiens, 1948, 1949).

7. This is from the unpublished transcript of a series of lectures given at the Committee on Social Thought at the University of Chicago in 1959.

8. I refer to a series of nine articles that appeared in the *Revue de l'Université d'Ottawa* in 1972: Vol. 41, No. 4; Vol. 42, Nos. 1 and 2. Francis L. Gammon, Jr., in "The Philosophical Thought of Yves R. Simon: A Brief Survey" is silent about *Prévoir et savoir*, 42, No. 2 (1972), 237–44. As a counterpoint to this, there is a work that shows Simon's influence and a utilization of *Prévoir et savoir:* John C. Cahalan's *Causal Realism: An Essay on Philosophical Method and the Foundations of Knowledge* (Lanham, Md.: University Press of America, 1985).

9. Trans. Victor M. Hamm (Milwaukee: Tower, 1942), p. 102. The original is *La marche à la délivrance* (New York: Éditions de la Maison Française, 1942), p. 10.

10. Alex Orenstein, *Willard Van Orman Quine* (Boston: Twayne, 1977), pp. 152–53.

11. In the unpublished correspondence, Jacques Maritain expressed some difficulty with the notion of philosophy within science. "I don't understand your philosophy within science very well. I rather think as Adler does that it is a kind of pre-philosophy of common sense on which science rests, but which does not enter into its texture." Jacques Maritain to Yves R. Simon, March 6, 1939. The vast manuscripts of transcribed correspondence between Yves R. Simon and Jacques Maritain (1929–1961) is housed at the Yves R. Simon Institute, Mishawaka, Indiana.

12. "Philosophy of Science," *Revue de Philosophie,* N.S. 6, No. 1 (January 1935), 59.

13. Ibid.

14. Yves R. Simon, "Causality and Indetermination," Lecture 16, February 26, 1959.

15. "Philosophy of Science," 62.

16. Simon to Maritain, October 6, 1944.

17. See reviews by Marcel Perrault, *Revue de l'Université d'Ottawa,* 15, No. 1 (1945), 248–49, and by Aron Gurwitsch, *Philosophy and Phenomenological Research,* 7, No. 3 (March 1947), 339–42. See references to critical realism in Yves R. Simon, *An Introduction to Metaphysics of Knowl-*

edge, trans. Vukan Kuic and Richard J. Thompson (New York: Fordham University Press, 1990), pp. 6, 22*n*32.

18. Review of *Prévoir et savoir*, 342.

19. "We prove that the principle thus put into play, the principle of lawfulness (*légalité*), is not enough, that science attempts equally to *explain* phenomena, and that this explanation consists in the identification of the antecedent and consequent." *Identity and Reality*, trans. Kate Loewenberg (New York: Dover, 1962), p. 10.

20. René Poirier says of Bachelard: "He thinks in opposition to his constant enemies: Meyerson, positivism, empiricism and, above all, in opposition to his pet hate, this poor realism." "Autour de Bachelard épistémologue," *Colloque de Cerisy-La-Salle sur le thème Gaston Bachelard, 1970* (Paris: Union Générale d'Editions, 1974), p. 17. The criticism of Meyerson is mentioned by most commentators on Bachelard. For instance, Jean-Claude Margolin, *Bachelard* (Paris: Seuil, 1974), p. 66, says Meyerson was the anti-model for Bachelard.

21. Yves R. Simon's editorial comment on the book is in the *Revue de Philosophie*, N.S. 8, No. 4 (July–August 1937), 355.

22. "With due allowance for the voluminous parts played by interpretations and constructions, the analysis of scientific facts brings forth a core of existential decisiveness by reason of which the fact can be called *the experimental absolute*. In this existential decisiveness, we recognize the crucial *difference* brought forth at the beginning of this paper, the immense qualitative distance between sensing an object and imagining it (no matter how vividly) or understanding it (no matter how clearly and distinctly), or constructing it (no matter how elaborately)." In *Philosophy of Knowledge: Selected Readings*, ed. Roland Houde and Joseph P. Mullally (Chicago: Lippincott, 1960), p. 95.

23. Ibid.

24. *Foresight and Understanding: An Inquiry into the Aims of Science* (New York: Harper Torchbook, 1963), p. 16. See also his *Return to Cosmology: Postmodern Science and the Theology of Nature* (Berkeley: University of California Press, 1982).

25. *Foresight and Understanding*, p. 23.

26. *Practical Knowledge*, ed. Robert J. Mulvaney (New York: Fordham University Press, 1991), p. 128.

27. Pp. 115–36. The essay originally appeared in *The New Scholasticism*, 27, No. 3 (July 1953), 280–304.

28. *Practical Knowledge*, p. 119.

29. Ibid., p. 128.

30. Ibid., p. 132.

31. Ed. Vukan Kuic (New York: Fordham University Press, 1986; repr. 1989), p. 94.

32. (Paris: Presses Universitaires de France, 1965).

33. (Cambridge: Cambridge University Press, 1984).

Acknowledgments

Originally published in Montreal in 1944, *Prévoir et savoir* (*Foresight and Knowledge*) is Yves R. Simon's most important contribution to the philosophy of science.

The author made a translation of chapter 1, and portions of his translation along with a translation of chapter 4 by Gerard J. Dalcourt were included in Yves R. Simon, *The Great Dialogue of Nature and Space* (Albany, N.Y.: Magi, 1970). As translators and editors we have been responsible for the Foreword and the remaining chapters (two, three, five, and six) as well as most of the notes in chapters 1 and 4. Earlier versions of the Foreword to *Foresight and Knowledge* and of chapter 2 appeared in *The American Catholic Philosophical Quarterly*, Summer, 1992; of Chapter 3, "The Vienna Circle," in *Semiotica*, 1994; of Chapter 5, "Some Remarks on the Object of Physical Knowledge," in *International Philosophical Quarterly*, September, 1992; and of Chapter 6, "Knowledge of the Soul," *The Thomist*, April, 1990.

The guiding principle of the edited translation has been to combine fidelity to the meaning of the text with flexibility as to expression. When the author has translated a part of his own book, the translator naturally feels some hesitation in making alterations. However, whatever initial reluctance there was to tamper much with chapter 1 was eventually overcome. We particularly thank John C. Cahalan for his meticulous reading of the first complete version of chapter 1 and we ended by incorporating many of the suggestions he made. These including restoring passages left out by Yves R. Simon as well as adding cross references to later works of the author. We are also grateful to Frederick Crosson, Gerard J. Dalcourt, Peter O'Reilly, Ralph McInerny, Dennis W. Moran, and George Stratigos for their critical comments and advice. Alasdair MacIntyre's recommendations and encouragement of this project were most appreciated. Our special thanks is due to Antony T. Sullivan, vice president of The Earhart Foundation, Ann Arbor, Michigan, for their support of this Yves R. Simon Institute project.

Finally, we should like to express our gratitude to Barbara Faria of the Department of Political Science and Diane Dupuis of the Word Processing Centre at the University of Windsor and to Roberta Ferkins, secretary of *The Review of Politics* at the University of Notre Dame, for their gracious help in the production of the book's manuscript.

Foreword

When the literature devoted to the problem of determinism and causality by our contemporaries is examined, it is impossible not to be struck by the contrasts existing between the clear and serene simplicity that characterizes the presentations of physicists, as long as they speak as physicists, and the nervous confusion that takes hold of minds once one moves from the level of scientific exposition to that of philosophic interpretation. There is nothing astonishing in the fact that philosophical conclusions are unable, in this domain as in others, to achieve the benefits of consensus; although philosophical certainty is of a demonstrative nature and, consequently, enjoys unlimited communicability in principle—a capability for intersubjective understanding—no philosophical doctrine, no matter how solidly established it may be, will ever be accepted in fact save by a small number of thinkers. The philosopher who is aware of his vocation should attach considerable importance to the distinction between being capable of intersubjective understanding in principle and being capable of it in fact. For a proposition to be capable of intersubjective understanding in principle, intrinsically susceptible of receiving a universal assent, it is enough that it be demonstrated, that is, joined by an obvious connection to the primary evidence of experience and reason. But the capacity for intersubjective understanding in fact, the property of actually rallying everyone's assent, or that of the greatest number, presupposes a host of conditions without any relation to the appropriate requirements of the capacity for intersubjective understanding in principle. In order that a proposition have the chance to become acceptable beyond the confines of a group of like-minded people, it is not enough that it be true, certain, and demonstrated; in addition it is necessary that understanding the demonstration require only commonly realized subjective conditions.

Numerous contemporary works in epistemology, especially those of the Vienna Circle, have shown that one of the constant characteristics of positive science is the search for a capability for intersubjective understanding in fact. In the definition of his own viewpoint, the positive scientist shows a concern to exclude as much as possible any consideration apt to make the actual meeting of minds exceedingly improbable. As a consequence of that, positive science gives up viewing and stating many aspects of the world, even among those that lend themselves to the establishment of certain knowledge. But there is a willing consent to this sacrifice as the indispensable condition of the greater advantages that the capability of intersubjective understanding in fact brings in the exercise of the social functions of scientific culture. The situation of the philosopher is completely different; nothing allows him to limit his object as a function of the necessities of that capability of intersubjective understanding in fact, and all of history is there to remind him that if he wants to be the one who achieves unanimous agreement, there is nothing left for him to do but to give up philosophy.

Many philosophers, bearing witness in this to a nonphilosophical spirit, cheerfully draw conclusions about a principle from a fact, and, under the pretext that actual unanimity is impossible in philosophy, give up trying to establish a system of truths capable in principle of imposing themselves on any mind by virtue of their intrinsic evidence. Thus, philosophy evolves toward literary expression, and the preference for such and such a system becomes a matter of intellectual, moral, and imaginative taste. In these conditions can one speak about a collaboration between philosophy and the sciences? The heterogeneity between science and what is called philosophy is too great, and one does not see how mutual understanding could be guaranteed. Generally, the scientist contemptuously shrinks back from a collaborator who he knows is not of one mind with him, and he decides to construct the philosophical interpretation of science himself. But then, whether he likes it or not, he borrows his principles of interpretation from the philosophies that dominate in a diffused state the environment in which he lives, and confusion triumphs, for nothing is more confusing than a philosophical idea divested of its technical precision and misfiled in a stock of "general culture," following a journey in which it has lost all its fine points in order to preserve only a largely unconscious and mainly destructive effectiveness. Let us note the example of the notions of causality, determinism, and chance. In the lamentable state in which these notions, burdened with incoherent meanings by incompatible philosophies and simplified by the work of popularizers, reach the scientist, they are all too often only the medium cause of disorder and ambiguity.

I will take the liberty of expressing some observations here about the situation of contemporary Thomism as it confronts problems in the philosophy of science. Many times it has been declared that the mere fact of having been established at a time in which positive science did not exist (or existed only in a rudimentary state) sufficed to prevent Thomism from becoming a truly modern philosophy, that is, capable of dealing with the problems of our time in a satisfactory way. This objection of principle has never touched serious minds very much, since it is just as evident that most of the great philosophical systems (materialism, idealism . . .) were similarly founded some centuries prior to the age of positive science. Several months before his death Émile Meyerson said to Father Rabeau, "I have nothing against your scholasticism; let it produce something in the philosophy of science, and we will think about it." Meyerson died too soon to be acquainted with *The Degrees of Knowledge*.[1] This book has shown that the purest and the most coherent Thomism, and the least inclined to accommodations, was capable of producing something great in the philosophy of science. *The Degrees of Knowledge* has the marks of a groundbreaking work. In spite of the bad historical times there is every reason to hope that Maritain's work will have been the starting point of a series of fruitful research projects.

But we should remember that the success of this research depends to a large degree on the way in which our efforts are received in the scientific world. Without the collaboration of the scientist, the philosophy of science cannot securely move beyond some statements of principle. The mutual trust between the Thomistic philosopher and the scientist is a basic condition of the progress of the Thomistic philosophy of science. Much effort has been expended with a view to establishing this mutual trust; it is appropriate to ask ourselves where we are now and what remains to be done.

A first point is established. We are present, on the side of Thomistic philosophers, at the decline of that spirit of *philosophic imperialism* which formerly was able to place the freedoms of positive science in jeopardy. Maritain's work has definitely liquidated the *tragic misunderstanding* begun during the Galilean–Cartesian age, and no one today any longer maintains that the very possibility of an experimental and mathematico-physical science is excluded by the principles of Aristotle's philosophy; but if there are Thomists who undertake to return this misunderstanding to the order of the day, let us dare to say that they are wasting their time.

A second point is established. Scientistic philosophy, by which I mean that mental conception that confers on positive science the characteristic of being the supreme or sovereign instance of wisdom, attributing to it a power of unappealable judgment over philosophy, conscience, and religion, has rather generally been discredited by scientists today.

A third established point is that in relation to the decline of scientism, on the one hand, and the progress in the history of ideas, on the other, the prejudice of *modernity* has lost some of its influence. In contrast to what occurred yesterday, numerous scientists today seem ready to recognize that the antiquity of a philosophy does not necessarily prove that this philosophy is worthless.

A fourth point, and this one is decisive, will be established the day when scientists will have understood to what extent Thomists share the same spirit as animates the man of science. Although Thomists are often obliged to talk like everyone else and to distinguish and even oppose the terms science and philosophy, and even though they in no way ignore the affinities that exist between and among philosophy, art, and religion, their conception of philosophy implies above all that philosophy is itself a science or, to be more precise, that the philosophic disciplines are each of them a science. We have had countless occasions to notice that the scientistic conception of philosophy is what contributes most to separating us from philosophers unfamiliar with our frame of mind and from an important segment of the cultured public. There are those who will not forgive us for this perpetual criticism of the meaning of words, these distinctions, this concern for precise definitions and exhaustive divisions, for rigorous deductions and necessary arguments, this horror of vague approximation. These are the admirers of what General Vouillemin so nicely called literary philosophy.[2] But all these mental traits that we owe to the Aristotelian school—the only school in which philosophical disciplines have been brought to a degree of technical elaboration comparable to that which accounts for the prestige of the exact sciences—constitute so many bonds of kinship with genuine scientists, even if it compels us to solitude as far as literary philosophers are concerned, a solitude in which we readily find consolation. Let genuine scientists be willing to carefully read our works with enough patience to overcome the inevitable difficulties of an unusual language; they will have no difficulty in recognizing that we belong to the same breed.

Notes

1. Jacques Maritain, *Distinguish to Unite, or, The Degrees of Knowledge*, trans. Gerald B. Phelan (New York: Scribner's, 1959).

2. In France a *docteur ès lettres* is the holder of a doctorate in literature or literary studies. Hence, the substitution plays on the title and indicates a literary philosophy [ed. note].

1
The Theory of Determinism

Chance

WHEN A PHYSICIST SPEAKS OF CHANCE, he generally has in mind events characterized by unpredictability. If he belongs to the old deterministic school, he quickly remarks that no events are unpredictable in themselves and that the notion of chance would be altogether meaningless to an intellect capable of grasping all the causes at work in nature.[1]

Thus reference to chance would signal our failure to establish the complete set of factors that control an event. Chance would not enjoy any real existence. It would be but a name given to a certain kind of human ignorance. Such a view is not only found among physicists, but has also been held by many theologians.

The problem of the universal predictability of physical events aside, let it be remarked that the definition of chance in terms of unpredictability sounds congenial to common sense. Indeed, it is in the spontaneous usage of common sense that the physicists found it. We often use interchangeably such expressions as "by chance" or "by sheer luck" or "unexpectedly." We describe as fortuitous those accidents that nobody could predict, which implies that predictable accidents are not fortuitous and cannot be interpreted as chance events. The following is an example of a predictable accident: from a window I watch two cars moving in perpendicular directions; these cars are at nearly the same distance from an intersection and their speeds seem to be almost equal. It is early in the morning, traffic signals are not yet on, and neither of the cars is slowing down. At this time of the day, drivers believe that they are alone on the road. I anticipate the collision, first as threatening, then as inescapable. A small fraction of a second before the accident, my prediction is extremely well established. Plainly, no one would say that these incautious drivers are victims of chance; rather, they are said to be the victims of their own carelessnesss because they knew, or should have known, what risk they were running as they approached an intersection without slowing down and watching. On the contrary, if a driver on a mountain road was caught in a landslide, everyone would say that that person was a victim of extraordinary bad luck.

In order to understand the philosophic significance of these interpretations, we must be aware that the expression "common sense" does not designate a function possessed of objective unity but rather an aggregate of notions that enjoy, on diverse grounds, the privilege of being accessible without any technical or scholastic training.[2] The unity of these quite heterogenous notions is not objective, but merely psychological.[3] Thus, common sense contains an elementary philosophy which is, in a certain way, the origin of all philosophy and of all science. It contains a tyrannical imagery that science and philosophy often have to fight; it also includes a practical vision of the world which, though right and sound in its own field—i.e., in the practical field—may hamper the perception of theoretical truth. To test the common sense interpretation of chance, let us consider an example devoid of human significance and free from the disturbances that originate in practical concerns.

Imagine two players separated by a screen and unaware of each other's presence. An observer sees one of them roll a ball at a target; the same observer sees the other player roll another ball at another target. The balls collide. In a second phase, the screen is removed, and the two are playing together. According to the rule of their game, they aim their balls so as to have them collide. Plainly, the first collision is a chance event, but it will not occur to anyone that in the second case the balls collided by chance. The second collision is not fortuitous, because we assumed an agreement between the players. They have unified their action. The collision of the balls was present in their intentions. The first collision was fortuitous, because the targets were diverse, the players unaware of each other's existence, and the actions nonunified. Predictability does not matter: the observer who twice saw balls rolling toward each other may have been able to predict the encounter as safely in the first case as in the second. The chance event is defined, not by unpredictability, but by the *nonunified plurality of the causal processes from which it results.*

This interpretation of the chance event as the result of an irreducible plurality of causes was expounded with remarkable lucidity by Augustin Cournot. In a celebrated page of the *Essay on the Foundation of Our Knowledge,* he points out that there exist in the universe of causes and effects *inter*dependent and *in*dependent series. "It is not impossible that an event occurring in China or Japan may have some influence upon events happening in Paris or in London. But, in general, it is certain that the program a Parisian lays out for his day will not be influenced in the slightest degree by what is then going on in some city of China in which Europeans have never set foot. These are like two little worlds in each of which series of causes and effects can be observed developing simultaneously which are not connected and which exercise no appreciable influence on one another."[4] The distinction between interdependent and independent series makes it possible to define chance as follows: "Events brought about by the combination or conjunction of other events which belong to independent series are called *fortuitous* events, or the results of *chance.*"[5]

Here is one of the examples given by Cournot. Two brothers who were serving in the same army unit were killed in the same battle. The conjunction of their kinship and of their misfortune may not be fortuitous. They may have been exposed to the same danger because they wished to stay close to each other. If, on the contrary, two brothers one of whom is serving on the northern border and the other in the South are killed the same day, there is every reason to consider this conjunction as a chance happening. Thus, when the two great brothers-in-arms, Desaix and Kléber, fell on the same day, indeed almost at the same moment, one on the field of battle at Marengo, the other, at the hand of a fanatic, in the city of Cairo, there certainly was no connection between the maneuvers of the armies on the plains of Piedmont and the causes that, on the same day, led the assassin to attempt his work. Furthermore, there was no connection between these diverse causes and the circumstances of previous campaigns along the Rhine which had led those interested in the glory of our arms to link together the names of Desaix and Kléber. The historian noting this odd coincidence might very well excite the surprise of the reader by saying that he can see in this event only a fortuitous happening, a pure effect of chance.[6]

We have suggested that the common sense interpretation of chance was governed by practical concerns. For the predictability of a fact, even though it does not alter the relation of this fact to its real causes, deeply modifies its human meaning. A predictable fact, whether fortuitous or not, is a thing that human prudence must take into account. We do not say that the careless drivers asked for their misfortune: one of them wanted to go to work and the other wanted to go home. But no human responsibility is involved in accidents that are humanly unpredictable, whereas we may be held responsible for accidents that we should have been able to predict. Common sense defines chance in relation to prediction, because it considers chance from the standpoint of action. We should not be surprised that philosophy has to correct common sense on a subject whose human significance never abates.[7]

Yet any discrepancy between the practical concept of a reality and its theoretical notion calls for explanation. If it is true that a predictable event which results from a nonunified plurality of causes remains a chance event in spite of its predictability, how is it that common sense attributes fortuitousness to unpredictable events alone? The answer is obvious: even though unpredictability does not pertain to the essence of chance,[8] chance events, by reason of their complexity and irregularity, are generally harder to predict than natural events. Inasmuch as a chance occurrence results from a nonunified plurality of causes, it cannot be predicted except on the basis of a complex system of initial data. On the other hand, to foresee that this bush will bear roses in the spring—except in the case of interference—all I need to know is that this bush is a rosebush. The irregularity of the chance event follows upon the nonunified plurality of the causes involved in its production; causes cannot operate jointly with regularity unless the regularity of their joint operation is guaranteed by the unity of a permanent principle.

The Philosophic Ambiguity of Determinism

Let us now try to see in what sense the notion of determinism and the notion of chance agree, and in what sense they exclude each other. It must be said, first of all, that the concept of an absolute chance, independent of all determinism, is a contradictory fiction. To see that contingency is grounded in necessity, it suffices to remark that the fortuitous encounter of two things in motion presupposes two things, each of which follows a nonfortuitous direction. The intersection of two lines presupposes two lines. A coincidence foreign to law and logic presupposes processes each of which obeys a law and has its own logic. Unless the philosophy of indeterminacy builds on an assertion of determinism, it is bound finally to deny the rationality of the universe and its reality.

The theory of chance not only presupposes the determination of the causal lines whose encounter constitutes the chance event, but also holds that the chance event, as soon as the causes from which it will result are actually posited, must occur inevitably, necessarily, and determinately. But this determination of the fortuitous is only *de facto* and not *de jure*. The necessity of the contingent is but an *inevitability*, a factual necessity; it is an *historical*, not an essential necessity. If you consider as factually existing, historically posited, an aggregate of causes whose actions converge without their convergence being embodied either in any of these causes or in the aggregate itself—which, by hypothesis, is nonunified—the result of their convergence, provided no free agent steps in, will occur necessarily (according to an historical necessity) without ceasing to be contingent and fortuitous: fortuitous, for it is not intended by any cause, or by any unified system of causes; and contingent, for it would not have occurred if another aggregate of causes had been factually posited.[9]

This notion of a merely historical necessity, or of a merely factual determinism, does not raise any difficulty so far as the demiurgical function of science is concerned.[10]

Engineers would not challenge the philosophy of contingency: they are interested in, say, regular telephone service between New York and San Francisco, and this is sufficiently guaranteed by historical necessity and factual determinism. But the theoretical ambition of science, its eagerness to explain, may bring about the denial of real contingencies and forcibly substitute a *de jure* determinism, an essential necessity, for merely historical necessities and merely factual determinations. For historical necessity is not explanatory, and there is the rub.

We now understand the meaning of the dialogue between Aristotelians and Stoics that St. Thomas recalls in his commentary on Aristotle's *On Interpretation*.[11] Aristotle does not say, as does one misleading formulation of the principle of causality, that all that happens has a cause and, consequently, is explained by a cause; he holds, rather, that the only being that has a cause is a *per se* being, that is, a being provided with essential unity. The Stoics, on the contrary, assert that everything has a cause, and although it may not be possible to explain an event through a single efficient principle, every event can be explained if a sufficient number of causes are brought to bear. These multiple causes, provided they are taken together, the Stoics held to contain the ground, the reason, the generating idea of the event to be explained. But this theory assumes a unified multiplicity and an explanatory causality. Plurality is not deemed irreducible; analysis, if carried far enough, will come upon causal unity.[12] Chance is eliminated, and its proper mystery disappears. So they conclude that all happens by necessity. The necessity spoken of is essential, and hence the event was written beforehand in the unity of a group of causes. This predetermination of the chance event is what Aristotle denies: "... not everything that happens has a [real] cause, but only *per se* being. What exists in merely accidental fashion does not have a cause, for, to speak properly, it is not a being. ..."[13]

If we have well understood what is implied by the irreducible plurality of the causes whose interference brings about the chance event, we shall not hesitate to say that an event brought about by chance is an event without a real cause. It does not have a real cause because it has several real causes whose plurality is irreducible. A thing cannot be a real cause unless it has real unity. Now, the plurality of the causes from which the chance event results has no unity except in our mind. We speak of it in the singular; we say *a* plurality, *an* aggregate, of causes, but it is the mind that supplies, here, a link that does not exist in the real world. The unity of this plurality or aggregate is a being of reason, just as much as a logical subject, a predicate, or an imaginary number. When chance is said to be a cause, let it be understood that the unity of this accidental cause is the work of reason.

To say that the principle of causality rules out the reality of chance involves the assumption that if chance were real its unity as a cause would have to be real. A being of reason—the unity of chance as a cause—is imaginatively projected into the real world and, when it is understood that it has no place there, chance is denied all reality. It follows from all this that a chance event is unexplainable, unintelligible, and nonrational. This is why rationalism is intent on denying the reality of chance. Every explanation consists in an identification. To explain the Pythagorean relation is to show that, within a definite system of postulates, there is a strict identity between what is called a right triangle and what is defined as subject of the property under consideration. To explain an event by its efficient cause is to show that the features of the event are, at least in part, identical with the features of the nature described as its cause. When there is a question of a chance event, no causal identification is possible.

Let us consider again the case of the uncautious drivers. One of them wanted to go to work, and the other wanted to go home. Suppose that no accident occurs. The first person actually reaches the place of work: this result is intelligible, explainable by reason of its identity with the representation that directed the driver and the car. The second driver actually gets home: this result is equally intelligible. But the collision does not resemble any causal antecedent, and it cannot be identified with the dynamism of any representation or nature. By reason of the very plurality of its principles, it is not identical with any of its principles. It involves an irreducible lack of intelligibility by reason of an irreducible lack of unity on the part of its causes. Hence, the rationalistic postulate of universal intelligibility, which requires the elimination of chance, requires also the elimination of plurality. Rationalism implies a theory of a universe that is *absolutely* one, fully revealed and fully actual. Diversity and change are reduced to mere appearances. But what becomes of the *real* universe if the reality of change and plurality is denied? Acosmism is the logical termination of the rationalistic endeavor. Thus, as Aristotle indicated, we would no longer be reasoning about things but about nothing.[14] Aristotle was referring to the monism of the Eleatics, but similarly Meyerson sees in the unitary and materialistic metaphysics of Parmenides the archetype of all attempts made at asserting the fundamental *identity* of things, be it at the expense of their *reality*.[15]

Yet it often seems, in our daily life, that we succeed in explaining facts that, according to the preceding description, should be considered chance events. If this is an illusion, our illusion still has to be accounted for. Let us try to analyze one of those chance events that everyone thinks he can explain.

On a winter day, an old man crosses a street on a steep hill. He slips on the ice. An automobile going down the hill cannot stop in time; the man is killed. His death is unquestionably a chance event, for, as Cournot would say, there is no connection, no interdependence, between the series of causes that determined the construction of the street on this hill, the series of causes that led the old man to cross this street at this very time, the series of causes that determined the formation of ice on this day and in this place, etc. But people keep on reasoning. For the family of the victim, the accident is easily explained; its cause is the obstinacy of the old gentleman. He had been warned that the pavement was slippery and that he would be wise to give up his daily walk. But he did not listen. For the driver of the car, the accident is also easy to explain. It had occurred to him that he might be unable to control his car in such weather on such a steep hill, but he was in a hurry and did not detour. He considers with remorse that his carelessness caused the death of a man. For the head of the Street Department also, the accident is clearly explained: an employee was ordered to spread ashes on slippery roads but thought that he could afford to skip this side street with little traffic.

But the meaning of these explanations is altogether practical. We would never have had the illusion that we could explain a chance event—by "explain" I mean to explain in the proper sense of the term, that is, theoretically—if we had from the beginning possessed a clear notion of what a practical explanation is. Theoretical thought endeavors to know what things are, and in order to know them perfectly, it needs to know them through their causes and in their causes, which is to explain them. Practical thought wants to know what we have to do in order to attain our ends, what we should have done in order not to fail of our ends, and what we must do in the future in order to stay along the path that leads to our ends. From the standpoint defined by the questions "What must we do?" "What should we have done?" "What shall we do?" to say that the accident is explained by the obstinacy of the old man, by the carelessness of the driver, or by the disobedience of the Street Department employee makes sense. But these explanations are totally meaningless with regard to the theoretical question of the nature of things and of the way in which they proceed from each other. From this theoretical standpoint, it would be necessary to choose among the heterogeneous and unrelated causes that the family of the victim, the conscience of the driver, and the head of the Street Department assign to the accident. Does it lie in the obstinacy of the old man, the carelessness of the driver, or the disobedience of the employee? It is impossible to make a choice. Each of these causes is a necessary condition of the accident and, so far as theoretical reason is concerned, no one stands out. It cannot be said that the accident is explained by the whole sum of its causes, for these causes do not constitute a unified whole. Again, the illusions of common sense are traceable to the fact that an insight whose significance is entirely practical is mistakenly applied to a theoretical problem.

Moreover, the illusion that a chance event can be explained is nicely balanced by another feature of common sense psychology: the feeling of admiration and awe that the contemplation of a fortuitous event brings about in us. The bewilderment of lovers thinking of the many encounters that were needed to bring about this masterpiece, their love, is an inexhaustible source of lyrical expression. Inasmuch as the causes of the fortuitous are several and are not unified, they explain nothing and leave us to our worries. Tired of seeking an explanation for the fortuitous event, the mind finds peace in the thought of Providence alone, and the bewilderment of chance, whether joyful or sad, often ends in an act of adoration. True, it is only on the level of the First Cause, on the level of the divine decree that organizes chance in a design whose ways are impenetrable, that the plurality of causes is finally unified. Should it be said that this final unification ultimately reduces the plurality of the causal lines, and that the acknowledgment of providential rule, as well as the assertion of universal necessity, amounts to denying the reality of chance? To this question it must be answered that the supreme distinction of divine government is precisely its ability to move creatures according to the mode of operation that befits them by reason of what they are. Under the influence of the First Cause, always infallibly effective, the natural event takes place naturally, the free event freely, the contingent event contingently, and the fortuitous event fortuitously. Concerning the real existence of chance, the problem is not whether the plurality of the causal lines is or is not finally unified, but whether this final unification takes place in nature or only in the Author of nature, who is transcendent to nature. The notion of natural necessity implies that the reduction of the causes involved in bringing about the fortuitous event occurs *within* the physical world. The reality of chance disappears. But the theory of providential government holds that this ultimate reduction takes place only on the level of the First Cause, and thus the fortuitous remains, and retains all its mystery, though it is no longer burdened with dread.[16]

Causality and Identity

Aristotle's starting point is, in a certain sense, directly the opposite of Parmenides'. The entire Aristotelian philosophy of nature rests upon the two facts that Parmenides is held to have denied: that of plurality and that of change. Yet Aristotle proposes to explain nature, and he would not question that an explanation is an identification. Because of serious shortcomings in Meyerson's own philosophy, his analysis may raise doubts; if explaining by identifying ends in monism and acosmism, must we not, in order to avoid these absurd consequences, conclude that explanation cannot be conceived after a pattern of identification? In our opinion, the error does not reside in the theory of explanation but in postulates relative to the intelligibility of the world. The world cannot be entirely reduced to unity, because it is but partly and relatively one. Being but partly and relatively one, it is but partly and relatively intelligible or explainable, for diversity bears witness to the presence of nonbeing in things. An explanation is indeed an identification, and when our discourse runs into an obviously irreducible diversity, we are given to understand that the aspect in which we are considering the world is really refractory to intelligibility.

The Aristotelian and Thomistic theory of causes, supplemented by the theory of virtual distinction,[17] has successfully defined the ways that reason must follow if it cares to respect the truly unexplainable aspects of the real without ever failing to explain what is truly explainable. At the very commencement of philosophic reflection upon nature, the conjoint fact of becoming and plurality runs against our eagerness to understand. But this mysterious universe is not in any way absurd. There is no absolute beginning. What is absolute does not begin, and what begins is not absolute. This is how the concept of origin forms in our minds. With beings that come into existence, the problem of origin is inescapable. To assume that such beings have no origin is to introduce contradiction into their very concept. What, then, is the origin of the being that comes into existence? It cannot be nothing, for from nothing nothing may come; and it cannot be being, for nothing becomes what it already is. Thus, we must conclude that its origin is intermediary between being and nothing, that it is a possibility of being, a being in potency, a potency distinct from nonentity and distinct from actuality.[18] To use the simplest of examples: consider the shape that clay receives under the hands of the potter. Where does this shape come from? Neither from the shape that clay possessed before, nor from nothingness, but from an ability to change shape, from an ability to receive the shape actually imposed by the artisan. The passive potency, the material cause, the passively indifferent principle of becoming, that which endures under changing forms, the permanent subject of transitory phases—such is the first factor of identity discovered by our mind in its search for unity. The explanation by material causality, the emphasis on a subject that remains identical under the successive multitude of acquired and lost determinations, is logically the first of all explanations; it is the most primitive of all, the easiest, and the one we are the most tempted to overuse. Materialism is the path of least resistance.

Let it be pointed out that material causality, such as it is understood by Aristotle, constitutes only a principle of relative identification. Many authors, among whom to our regret we have to number Meyerson, conceive the potency to be this or that, as an actual preformation of this or that, as a hidden act.[19] This is a radical misinterpretation of Aristotelianism. The being that is this or that in potency alone, absolutely speaking, *is not* this or that. It is able to become this or that, no more. Becoming brings genuine novelty. The continuity provided by the material cause is merely that of a link and does not suppress the heterogeneity of the phases that it connects with each other.

Moreover, the notions of material causality and passive potency would make no sense if they were not immediately related to the notion of active power or efficient causality, which expresses a principle of heterogeneity. The being that becomes, the being that moves from the possibility of being such and such to the act of being such and such, undergoes this transition by virtue of an agent distinct from itself. Whatever is in motion is moved by something else. This formulation of the principle of causality is essentially dualistic. It emphasizes the word *other* and can be described as *the principle of the otherness of the cause*. Whenever a cause is needed, identity is negated. Every causal relation implies the duality of the cause and the caused. In absolute identity, that is, in God, there is no causal relation. God is not cause of himself but *is* without a cause. Let us become aware of the paradox involved in these familiar remarks.

When we explain an effect by its cause, we propose to manifest a link, some sort of continuity, an aspect of identity, rather than to emphasize what is irreducible in the diversity of the things. Why is it, then, that in its first formulation the principle of explanation by efficient causality stresses *otherness*? Why this emphasis on the *other* (*ab* alio *movetur*)? The answer is found in the analogical conception of being as unity and multiplicity indissolubly bound up to each other, supporting and conditioning each other.[20] By relating an effect to its efficient cause, a later phase to a prior one, we truly propose to place in evidence a connection and a continuity, and thus to achieve an identification. But an acknowledgment of otherness and an assertion of diversity constitute the *sine qua non* of such an identification, connection, and continuity. Let us state once again the problem of the origin of the being that comes into existence; by declaring that it comes from being in potency, we obtain an answer that is but partial and would be unsatisfactory if it were not quickly supplemented. Actual being, indeed, is more of a being than being in potency. From actual being, let us subtract potential being. There remains a plus, and this plus is precisely the actuality of being. But what is the origin of this ontological plus, of this actuality of actual being? To place the origin of this plus in potential being, which is less than actual being, is to trace to nothing all the difference existing between the more and the less, between the state of actuality and the state of mere potentiality. Thus, either the actuality of the being that becomes actual proceeds from nothing, or it proceeds from a being distinct from the being that becomes. To say that it proceeds from nothing is nonsense, and to say that it does not proceed at all amounts to giving up identification and explanation. Without a transition of the same to the other, becoming would be reduced to a succession of sames without any relation to each other. In order to safeguard unity, we assert diversity. In order to assign an origin to the actuality of being, we trace this actuality to another and pre-existent being. The active power to which the realization of the passive potency is traced actually possesses the ontological plus that distinguishes actual from potential being.[21]

This formulation of the principle of causality, "whatever is in motion is moved by something else," expresses only the indeterminate existence of a link between the same and the other. It demands to be completed by the formulation of the principle of proper causality, "every agent produces something similar to itself,"[22] and by that of the principle of finality, "every agent acts for the sake of an end." These two principles express the predetermination of the effect in the cause and constitute the immediate foundation of the concept of determinism.

Let it be said that in every cause really distinct from its action a twofold predetermination of action and effect is required: a formal predetermination (directive idea) and an existential predetermination (tendency, finality).[23] It is obviously impossible to achieve any kind of work unless one bears in mind an idea or a pattern, no matter how vague, of the work to be achieved. It is equally obvious that the realization of an idea presupposes a leaning, a tendency, an attraction, a love, a will. Let us consider an action at the time when it emanates from the agent and produces its effect. It cannot emanate unless it is determinate. An action describable as *any* action cannot emanate any more than a being describable as *any* being can exist. But where would the quality of the action—its being such and such—come from if it did not come from the agent? And how would it come from the agent if it were not, in the agent, pre-existent to action? All this is so clear that no one hesitates to infer the nature of things from the characteristics of their operations. Operations are the expression of pre-existent natures. Everyone calls the person who acts justly an honest person, and the person who plays the violin beautifully a musician.

As to finality, which is a predetermination to acting rather than not acting—whereas the leading idea is a predetermination to acting this way rather than in another way—we assert it virtually whenever we assert the reality of becoming and, on a deeper level, that of universal activity. If it is true that the world moves, that things are acting and reacting rather than motionless and dead, there must be within them, prior to action, a principle that causes them to act rather than to be inactive. If becoming is real, if the passage of time brings genuine novelty, if the future is not reducible to the past and the past to an everlasting present, there must be in the present (loaded with the past) a tendency to engender the future, a tendency that perpetually drives the present into the past. The theory of finality implies no more than this; it does not by any means involve an animistic, hylozoic, or panpsychistic theory of universal motion.[24] An idea that governs an activity and a tendency to act actually are not necessarily psychological entities. In order to account for an organic reflex, it is not necessary to posit representations and emotions in the muscles and in the medulla; the nature of the organs suffices. But whether nature itself, as source of regular activity, as *idea* and as *tendency*, ultimately participates in an idea that would be subsistent thought and in a will that would be divine government is a question that does not belong to the present inquiry.[25]

Whereas the first formulation of the principle of causality emphasizes the real distinction between the cause and the caused, the principle of the predetermination of the effect in the cause emphasizes their similarity. The efficient cause and its effect resemble each other. They are identifiable in part. They are identifiable inasmuch as the essential features of the cause are recognizable in the effect. The concept of proper cause is now quite clear. Buffon used to say that his servant Joseph, having awakened him very early every day, was responsible for several volumes of the *Natural History*.[26] But these volumes did not pre-exist in the brain of the valet; they were conceived by Buffon. A proper cause contains the likeness of the effect; the accidental cause does not. The latter is not explanatory, because it does not allow for any identification. If we want to explain the *Natural History*, we shall inquire into the readings of Buffon and his observations. There we have a good chance of finding a content like that of the *Natural History*. Joseph will not be mentioned.

The Concept of Proper Causality in Non-Ontological Knowledge[27]

But why should this example be borrowed from the order of human activity? Some critics will hold that it would be hard to find a clear example of proper causality in positive science. At this point, a consistent epistemological approach takes shape. Everyone knows that positivistic philosophy gives up the explanation of things and describes science as essentially dedicated to prediction. We have shown that among the realities that can be termed the efficient causes of an event the proper cause alone is explanatory. But if we must concede that positive science, by reasons of its own object and methods, is unable to disclose relations of proper causality, it is necessary to conclude that it must give up explanation. Now, the writings of scientists and epistemologists are crowded with observations designed to show that positive science constantly tends to eliminate proper causes and to replace them with such substitutes as the "total sum of necessary and sufficient conditions." We are, here, touching upon one of the most obscure problems of epistemology. First of all, let it be remarked that the concept of cause, inasmuch as it is related to the being of things, is an ontological concept. Such a concept operates freely, uninhibitedly, without any particular check or restriction in the domain of ontological thought, that is, in common sense and in philosophy. But in a system describable as non-ontological, or even anti-ontological, such as the positive science of nature, the concept of cause labors under restrictions and undergoes transformations parallel to those undergone by the concept of being itself. The question is whether these transformations go so far as to destroy the concept of proper cause and consequently to deprive positive science of all explanatory significance.

The only satisfactory answer would consist in an exposition of the diverse transformations that the concept of causality undergoes when the mind moves from the ontological to the positive level. These transformations are diverse, for positive science uses a multiplicity of methods. The following remarks may give an idea of what such a systematic exposition would be like.

(*a*) Our first remark will be derived from a distinguished field of positive knowledge, that of psychology. In the field of psychology, the ideas governing activity, the forms communicated by the proper cause to its effect, are reducible to representations. In the most favorable cases, at least, such forms can be known from within and communicated to the observer by the words of the observed subject. Thus, we are able to ascertain their likeness with the event to be explained. If Freud appears to us as a very great scientific mind, it is above all because of his determination to find, at the cost of laborious investigations in the darkness of the unconscious,[28] psychological forces that resemble the behaviors that are to be accounted for. Let us take an example from the slips that popularizers have made familiar. When a gentleman says: "Dann sind Tatsachen zum Vorschwein gekommen" instead of saying "zum Vorschein," we consider that the sound "schwein" is explained when we come to realize that this gentleman inwardly declared that these "Tatsachen" were "Schweinereien."[29] It would be easy to show that sociological inquiries inspired by the notion of *comprehensive interpretation*[30] take advantage of the particular facility that we have to ascertain the resemblance existent between cause and effect when the cause or, more precisely, its leading form is a representation.

(*b*) Second remark. In such a discipline as anatomy, where observation, induction, and sheer description obviously play a predominant part, the mind remains constantly guided by the resemblance of the forms involved in causal relations. A philosopher had enrolled at the school of medicine; confronted by the obligation of remembering the meticulous description of the humerus, he could not help feeling out of his depth. Here an eminence (or bump) supports a muscle attachment; there a groove provides passage for a neural course. At first no intelligible relation lightens the burden inflicted upon memory. One must confess that things are such and be sure not to forget it. But when one has studied the whole anatomy of the arm—bones, muscles, vessels, and nerves—the arrangement of the parts becomes intelligible in relation to the proper effect of the whole. In the anatomy of the arm, one can recognize the outline of such movements as grasping and gesticulating. The curve described by my hand when I point to an object placed in front of me implies a set of contractions which requires a system of muscles, nerves, and vessels arranged in conformity with the curve to be described. Thus, we come to understand why there must be a muscle attachment here and a groove facilitating the passage of a nerve there. And simultaneously we understand that the famous formulation, science knows the *how* of phenomena but not their *why,* is uncertain and quite uselessly cumbersome.

(c) Third remark. Consider the formula of a chemical reaction. Between the symbols to the left of the sign of identification and those to the right, likeness is obvious. They are the same elementary symbols. Their arrangement alone differs. In what order of causality does the identification take place? Primarily in the order of material causality. The sign $\to$ in the equation $H_2 + Cl_2 \to$ signifies primarily that the elements H and Cl, which exist in a certain state as bearers of a certain set of properties prior to the reaction, continue to exist after the reaction, though in another state and as bearers of another set of properties. The identification expressed here, which confers upon chemical studies such an inebriating intelligibility, concerns, above all, the permanent subject of the transformations, namely, the set of elements symbolized. But this is not all. We consider the compound as the effect or result of the mutual action of the components. The expression re-*action* testifies that the concept of efficient causality is present in our minds. If we undertake to determine the cause of this reaction, we shall mention numerous and disparate principles. We shall say that the elements cause the compound through their mutual action. We shall also say that light, because the reaction does not take place in complete darkness, must be counted among its causes. We shall also mention the bringing together of the chemicals, and ultimately we may say that only the whole set of the necessary and sufficient conditions should be described as the cause of the phenomenon. All that sounds perfectly reasonable indeed, but it must be pointed out that among the conditions of the phenomenon there is one that enjoys a distinguished character; it is the *nature* of the reacting elements. We certainly can say that light causes hydrochloric acid, but light does not cause it in the same way as chloride and hydrogen do; light causes it less properly. We believe that the elements are causes of the compound in a more proper sense than light is because we perceive a resemblance between the elements and the compound, whereas no such resemblance appears between the compound and light.[31] Not only does the identification $H_2 + Cl_2 \to 2HCl$ express the permanence of a material cause; it expresses also, though with less emphasis, the resemblance existing between the distinguished efficient cause, $H_2 + Cl_2$ and the effect, 2HCl.

(*d*) Fourth remark. In the more highly mathematized parts of the science of nature, the concept of efficient causality goes through strange adventures. Let us content ourselves with suggesting that insofar as causality is reduced to a constant relation between two magnitudes one of which is conceived as a variable and the other as a function, what disappears is the concept of production, of realization, of activity, rather than the concept of *proper* causality as distinguished from accidental causality. Consider the Newtonian equation

$$F = K \frac{mM}{r^2}$$

Between the terms of the formula the relation of efficiency has disappeared, but not the relation of resemblance. Between F and mM there is a resemblance since the increase of these magnitudes is proportional; between F and r^2 there is an inverted resemblance. At this point the question is where an equation such as

$$F = K \frac{mM}{r^2}$$

can be understood in a genuinely physical and not merely mathematical sense, if it is understood without any reference to the concept of a realizing power, of an activity, and of a communication. The least that can be said is that for the most daringly mathematized physics, the complete exclusion of the concept of activity is but an unattainable limit.[32] Now, if what is eliminated is the relation of causality rather than the relation of resemblance, it must be said that in proportion as physics retains the concept of efficient cause it understands the latter as the bearer of features whose resemblance is found in the effect—in other words, as a proper cause.

The Scientific Ideal and the Real World

This brief inquiry into causality leads us to further elaboration on the theory of determinism. Let us imagine a universe strictly unified, that is, consisting in a single proper cause that would contain within itself all the physical conditions of its operations. Should it be said that the positing of such a cause entails with absolute necessity the emanation of its effect? Yes, if the standpoint is physical; no, if it is metaphysical. If our scope is limited to the order of physical causality, an isolated cause cannot be prevented from eliciting its effect, because by hypothesis no conflicting principle exists beside this cause. But if our scope is metaphysical, the necessity that governs the occurrence of the effect is no more than conditional; for a secondary cause, even though solitary, can be impeded by the suspension of the divine concursus, which is necessarily required for any exercise of created causality. Every metaphysician will appreciate the significance of this remark, which sharply reminds us that existential states are foreign to the domain of scientific knowledge. Let us take the term "science" in its most profound and rigorous meaning, according as it designates a system of theoretical knowledge exclusively and indefectibly relative to the true. In order for theoretical knowledge to be absolutely free from error, its object must not admit of being otherwise than it is. It must be necessary. But there is only one necessary existence, namely, divine existence; no created being implies its own existence. Because divine existence alone is identical with the essence of which it is the act, it is the only existential state that bears the character of a scientific object.

It is hardly necessary to recall that our factual knowledge, that is, what is taught as science in our courses and treatises, contains only a small nucleus of truly scientific propositions, that is, of propositions expressing in complete evidence absolutely necessary relations. A nucleus of genuinely scientific certainties animates and sets in order *a variegated ensemble, in which merely factual assertions, opinions, and beliefs teem.* This ensemble is what we improperly call "science" when we consider what the human mind has actually constructed in its endeavor to construct science. But the improper meaning of a term is not intelligible save in regard to its proper meaning. A defective realization is not intelligible save in regard to a faultless ideal. The movement and the data of historically existing science are not intelligible save in regard to the ideal of a knowledge that would exclude from its object any element of contingency and clearly read the *necessary formulations of essential possibilities.* Whereas every created existence is altogether contingent, the possibility of any possible thing is as necessary as the being of God. The necessity that grounds the constitution of the possible is identical with the necessity of the divine being. Hence, the extraordinary fascination that the ideal of theoretical science exerts upon our intellects as soon as we understand the sublime requirements of this ideal. Whenever we succeed in establishing a truly scientific proposition, we are delivered to an eternal truth and become one with an aspect of subsistent eternity.[33]

On the level of abstract possibility, to say that A is cause of B is to say that the ability to bring about B is a property of the essence A. Between this essence and this dynamic property there is no real distinction. Between the essential tendency of a thing to cause such and such a proper effect and the essence of that thing, there is only a virtual distinction. It is more than a distinction between words or signs; it is less than a distinction between things or real modes; it is a distinction between objective aspects of one and the same thing.[34] Causal identification attains its perfection, without change being stopped or nature destroyed, when we succeed in seeing that between the essence of a certain thing and its tendency to bring about such and such an effect there is real identity, when we see that this essence and this tendency are but two distinct aspects of one and the same thing.[35] True, such rational penetration takes place only in a small number of distinguished cases; but this triumph of our scientific ideal, no matter how rare, renders meaningful all our endeavor to understand nature.

We have seen that even in the hypothesis the most favorable to the requirements of causal explanation, namely, that of a universe reduced to a single cause, containing within itself all the physical conditions of its operation, it would remain false to say *in a metaphysical sense* that the positing of the cause necessarily entails the positing of the effect. There always may be impediment on the part of the first cause, and the most simple universe, if it is considered as actually existent and active, is separated from the scientific universe by the properly infinite distance that takes place between the negation and the assertion of contingency. But we know that the real world is diversified. Even if the difficulties resulting from a nominalistic theory of universals are supposedly removed, it remains that physical species are several and that within each of them the multitude of individual existences admits of no *de jure* limitation. With regard to causality and chance, the plurality of individuals plays a more decisive role than the plurality of the species, for the specific unity of a multitude of individuals does not prevent the latter from encountering each other in their activity and bringing about fortuitous results. A glance at individual unity will help to understand some aspects of the problem of determinism.

Let us recall, first, that the law of individuality is absolutely universal. Outside the mind, nothing can exist save the individual and the collective entity made of individuals. Many thinkers hold that individuality is a proper characteristic of the living. Indeed, every transcendental property of being reaches a higher degree of intensity on the higher levels of being; the individuality of a living thing is more of an individuality than that of an inanimate thing, just as the goodness of a free will is more of a goodness than that of a thing of nature. But most of all, the discrimination of the individual, and the precise determination of its boundaries, are incomparably easier in the case of the living—especially the higher living organisms—than in the case of the inanimate body. Our discussions of individuation constantly use examples from the world of life. This does not mean that we question the reality of individual units among the nonliving; it merely means that we fall short of certainty when there is a question of designating such units. We count the sheep in a herd and say without fear of error that the number of individual units contained in the herd equals the number of the sheep. But how would it be possible to count the individual units contained in this wooden table or in this block of granite? Common thought wisely avoids this question, and is satisfied with an altogether practical notion of individuality: in a purely practical sense, two stones are two individuals. Philosophy asserts the existence of the inanimate individual, but, so long as it is restricted to data supplied by common experience, it has nothing to say about its boundaries. New possibilities are opened when scientific experience supplies philosophy with more refined data. Scientific experience, being microscopic, is better adapted to the small dimensions of the inanimate individual. It is not exceedingly daring to conjecture that the inanimate individual coincides in some cases with the crystal, in some cases with the molecule, in some cases with the atom, and in some cases with elementary particles. It is reasonable to hope that conjectures of such a kind will gain in assurance and in precision as an effect of progress in scientific investigation and in the philosophic interpretation of scientific data.

Now, if it is true that chance results from plurality, the field of chance grows as discrimination of plurality becomes keener. As we become able to perceive a greater number of factors, the part played by chance appears correspondingly greater. The microscopic approach to inorganic individuality was bound to multiply the conceivable interferences. The transition from the macrophysical to the microphysical standpoint was bound to be accompanied by a promotion of chance, inasmuch as the mind was now confronted with a *multiplicity* unknown to the rough insights of common thought and of classical physics.

And that in fact is what has happened. In a paper composed at the beginning of this century, Henri Poincaré wrote:

> We cannot foresee in what way we are about to expand; perhaps it is the kinetic theory of gases which is about to undergo development and serve as model to the others. Then the facts which first appeared to us as simple thereafter would be merely resultants of a very great number of elementary facts which only the laws of chance would make cooperate for a common end. Physical law would then assume an entirely new aspect; it would no longer be solely a differential equation; it would take the character of a statistical law.[36]

Thus, the transition from the "simple" to the "very great number" brings about the substitution of statistical laws for laws of causal description.[37] Now, the integration of chance is what basically characterizes statistical law. For the first-year student who repeats Boyle's experiment,[38] a gaseous mass has the character of an individual entity; one does not ask whether it resolves into a multiplicity of small masses each of which would act on its own as the spectators at a theater do in case of fire. It is postulated that this gaseous mass acts as if it were one subject of existence and activity. Is chance thereby denied? By no means. The very formula of Boyle's law refers to possible disturbances that the experimenter must prevent, for instance, temperature must remain the same, for a change in temperature would alter the result. But the so-called causal law, as it warns us against the substitution of a chance event for a process of proper causality, postulates that the elimination of disturbing factors is theoretically possible. On the contrary, if we consider that the undivided subject of existence and activity, that is, the individual, is not the gaseous mass enclosed in this container, but a molecule so small that it would fit in any manageable container a billion times over, we no longer can study anything else than the general effect of billions of actions and interactions. Chance is now integrated in the law.

Many thinkers have judged that this integration of chance in law constituted the most radical of all the revolutions ever undergone by the scientific mind and marked the definitive invalidation of the regulating ideal that science received from Greek rationalism. This ideal was briefly described above; this description must now be reconsidered and supplemented.

At the dawn of Greek thought, the philosophy of universal mobility is represented by the school of Heraclitus and Cratylus. The universe of sense experience is a flux where natures lose their identity, for they do not contain any principle of stable existence. Surrendering to the deceptions of a phenomenal universe where everything is in motion, early Greek thought goes through a gloomy phase. The phenomenon, inasmuch as it is carried by universal becoming, appears as the enemy of science. For Plato, a pupil of Cratylus, the scientific object must not be sought in the world of sense experience but in a supra-sensible universe of ideas and numbers. The phenomenal universe proves inconsistent as soon as it is subjected to theoretical analysis. It is merely an object of opinion. Whatever cognition we have of it is uncertain and contradictory; its real value is only practical. The scientific outlook implies first of all a conversion to the universe of things intelligible, and this conversion—here is the decisive mark of eternal Platonism—implies aversion to the flux of sense appearances. A science of nature is impossible.[39]

The point of departure for Aristotle is none other than the Heraclitean universe in perpetual change. But in the universal flux Aristotle discerns some islets of stability, some principles of order; these are the universal natures of things, the types, the laws of action, of passion and of interaction. It is very true that all things change unceasingly, if you understand, by saying all things, all such individual realities as are apparent to the senses. But within the changing individuals there is a universal type which does not change. In regard to the sensible world, the scientific mind ceases to observe an attitude of aversion to adopt an attitude of abstraction. The phenomenon becomes a friend of science; it leads the mind to the foundations of things. The phenomenal regularities, the constant relations between phenomena—in short, *the apparent order within the observable world*—reveal the intelligible order in which the mind satisfies, at least in the most favorable cases, its urge toward certainty and clarity. The leading idea of this philosophy of science can be expressed as follows: the chance event, inasmuch as it is not predetermined in any principle of activity, has the law against it and, consequently, must have the character of an exception; the natural event, inasmuch as it results from the undisturbed operation of a proper cause, has the law behind it and, consequently, must enjoy frequency. Existential and phenomenal regularities reveal the nature of things and lead the intellect to the perception of the eternal necessities of the possibilities, where the cause is identified with the thing and where the tendency to bring about a certain effect proves identical with a certain essential constitution, a definite form of being. Thus, in the line of discovery, that is, in the analytical or ascending line of science in the making, the apparent order—that is, the observable regularity—leads to the intelligible order. Correspondingly, in the descending and synthetic way, which is that of established science, the intelligible order—that is, the necessary relations of abstract possibilities—accounts for the apparent order. The mind moves from order to order. It designates, as the source of the order that is first for us, the phenomenal regularities; as the order that is first in itself, the necessary constitution of possible beings.

Henri Poincaré refers to this concept of nature and of scientific explanation when he writes, in somewhat awkward terms, that the Law in the eyes of the ancient thinkers was an internal harmony. "It was either an immutable type fixed once for all, or an ideal to which the world sought to approach."[40]

Poincaré immediately remarks: "Newton has shown us that a law is only a necessary relation between the present state of the world and its immediately subsequent state. All the other laws since discovered are nothing else; they are in sum differential equations. . . ."[41] Newton would have been the first to work out this ideal form of the physical law.

It would be extremely interesting to inquire by what logical process and through what historical phases the Newtonian conception of law was substituted for the Aristotelian conception. Let us only remark that for Newtonian science, as for Aristotelian, the apparent order of the universe, namely, phenomenal regularities, originates in an intelligible order. Minds trained in Newtonian disciplines were never tempted to place anything else than order at the root of order. This is established by the metapositive—nay, metaphysical—character of determinism in the works of classical physicists. It is a certain vision of the intelligible order which commonly led the physicists of the classical period to deny chance, contingency, and freedom, as incompatible with scientific determinism. The permanence of the Parmenidean myth in mechanistic systematizations expresses dedication to a basic pattern of intelligibility.

As a result of a microphysical approach, statistical formulas have been substituted for causal formulas, in an increasingly large domain of scientific inquiries, and the temptation to place something other than order at the root of order has been felt by many minds. From innumerable writings on this subject let us quote, because, of its clarity, an essay by Erwin Schrödinger.[42] This scientist remarks that Hume, in his celebrated discussion on causality, did not deny that some regularity obtains in the universe; but the very existence of such a regularity has been doubted in the last few years:

> The basis of this skepticism is the altered viewpoint which we have been compelled to adopt. We have learned to look upon the overwhelming majority of physical and chemical processes as mass phenomena produced by an immensely large number of single individual entities which we call atoms and electrons and molecules.... The exact laws which we observe are "statistical laws." In each mass phenomenon these laws appear all the more clearly, the greater the number of individuals that coöperate in the phenomenon. And the statistical laws are even more clearly manifested when the behavior of each individual entity is *not* strictly determined, but conditioned only by chance.... If an initial state, which may be called the cause, entails a subsequent state, which may be called its effect, the latter, according to the teaching of molecular physics, is always the more haphazard or less orderly one. It is, moreover, precisely the state which can be anticipated with overwhelming probability provided it is admitted that the behavior of the single molecule is absolutely haphazard. And so we have the paradox that, from the point of view of the physicist, chance lies at the root of causality.[43]

Schrödinger illustrates his conception with two examples. The first is imaginary. A well-ordered library is open to ill-mannered people who take the books out of their places and leave them all over the place. At the beginning of this fictitious observation, the eighty volumes of the works of Goethe are placed next to each other; after a week, twenty of these volumes are scattered, sixty are still in the right place; at the end of the following week, fifteen more have left the shelves; at the end of the third week, eleven others. "The number of volumes in their proper positions diminishes in accordance with the exponential law, or Law of Geometrical Progression, as the mathematicians call it."[44] The other illustration is the well-known example of life insurance: A company calculates with remarkable accuracy the percentage of deaths among its policyholders in a defined period, although it is impossible to foretell the future of any individual policyholder.

Schrödinger says that two interpretations are possible. It may be held that statistics really expresses the essence of the laws of nature. If such is the case, the concept of causal connection must be rejected. "We shall be especially inclined to sacrifice the causal principle if we follow Hume in recognizing that it is not a necessary feature of our thought, but only a convenient habit, generated by the observation of that regularity in the course of actual occurrences the merely statistical character of which is now clearly perceived."[45] But no experiment forbids anybody to prefer the opposite interpretation, according to which the behavior of each individual atom is determined by "rigid" causality, and statistical regularity is an ultimate result of the determination of elementary processes. The first position is "revolutionary" and implies that chance is primitive; the second is "conservative" and implies that chance is subjective. Schrödinger seems to think that it is up to philosophy to choose between the two.[46]

The "conservative" position, Schrödinger says, will be preferred by those who are not convinced by the critique of Hume and hold that the principle of causality is a necessary law of thought. Let us say that it will also be held, with greater force, by those who see in the principle of causality a necessary law of being and do not feel bound thereby to attribute to chance a merely subjective signification. The statistical law would express a state of affairs resulting from processes that can be roughly outlined as follows:

1st phase. A determined elementary movement, a natural event, not a chance event.

2nd phase. Innumerable encounters of individual causal lines, chance.

3rd phase. Balancing of individual irregularities in the regularity of an overall effect.

Obviously the main difficulty concerns the transition from the second to the third phase. This difficulty grows when one remarks, as Schrödinger insistently does, that the mass effect will be the more regular as individual behaviors are more completely irregular.

At this point, the decisive issue seems to be the *physical,* as opposed to the mathematical, meaning of probability. The notion of mathematical probability is established as follows: Let us imagine a lottery in which a thousand tickets are sold; only one will win; my chance of winning is proportional to the number of tickets I own (favorable cases, f). Let us now imagine two lotteries, one of which sells a thousand tickets and the other a hundred. A gambler owns one ticket in each lottery: his chance of winning is ten times greater in the second than in the first. The probability of winning is inversely proportional to the total number of the possible cases, s. If a gambler buys all the tickets ($f = s$), probability becomes certainty. As for the onlooker who has no tickets ($f = 0$), he is also certain, but certain not to win. Mathematical probability, $p,$ is equal to the ratio

$$\frac{f}{s}$$

of the favorable cases to the total number of possible cases. But we are obviously assuming that all possible cases have an equal chance to be realized, in other words, that they are equally likely or equi-probable. The definition of probability implies the concept of equi-probability or equal probability, and it seems that we are caught in a vicious circle. The circle can be broken, however, by considering that the equal probability presupposed by the definition is *physical,* whereas the probability defined is *mathematical.* It is impossible to speak of mathematical probability except in relation to events whose physical predetermination is assured in the same degree. As I leave for the university, I intend to be back home about tea-time. Four cases are conceivable: either I shall actually be home about that time, or I shall be held up by some chatterbox, or I shall be the victim of an accident and either taken injured to the hospital or taken home dead. Let the uneventful homecoming be considered the favorable case. No one will pretend that this case has only one chance in four to be realized. Why are we forbidden to use here, without any further elaboration, the formula

$$p = \frac{f}{s}?$$

Because the factors capable of bringing about each of the four events are not equally assured I am not disposed to let myself be held up by chatterboxes; my sight is good, my hearing is fine, I am well disciplined, and I do not cross streets against the traffic lights. If I were interested in gossip, absent-minded, color-blind, and hard of hearing, the physical probability of the four possible events would be closer to equal, and my family would worry every time I go out by myself.

On the basis of physical equi-probability, we infer the actual reign of mathematical probability. Consider a regular die: the probability of any one side coming up is held to be

$$\frac{1}{6}.$$

If this ratio is not realized in a long series of rolls, we shall conclude that there is a hidden irregularity. "This prediction," Guido Castelnuovo writes, ". . . is not justified by any rigorously logical method, but it satisfies a rational requirement of our thought which we express by the principle of sufficient reason. It is assumed that to a diversity of cases that are equally conceivable there must correspond an equal distribution of actually observable cases, so that the symmetry of causes be matched by symmetry in the effects. Such a prediction may well be given the lie by experience, but then our doubts will bear on the regularity of the die used."[47] We believe, on the contrary, that the said prediction, insofar as the principle of rationality is involved, proceeds from a rigorous logic. Given the physical equi-probability of the possible events, the frequency of the coming up of any particular side is not determined by any physical cause, as would be the case if the die were loaded, and can be determined only by mathematical form, namely, the ratio of number 1 to the total number of the possible cases.

But the physical equi-probability of the possible cases does not presuppose only the regularity of the die; it also requires disorderliness in the successive acts of rolling. If the die is cast by a perfect machine and if its initial position is always the same, it will always fall on the same side, equi-probability will be disrupted, there will no longer be probability, but certainty. One side will have a monopoly on the chances, like the lover of a sure thing who no longer gambles since he has bought all the tickets for a lottery. Only an ill-assured cause can give all possible contingencies an equal physical assurance.

In fact, we notice that one side of a six-sided die comes up about one time in six. Between the *a priori* probability of the uncertain event

$$(p = \frac{1}{6})$$

and the actual ratio of the number of favorable cases—for instance, the coming up of this particular side—to the total number of cases, there is a discrepancy whose absolute value increases and whose relative value decreases as the series of rolls goes on. The triumph of mathematical probability is the more complete as the number of rolls is greater; this is because the multiplication of the rolls procures a more complete physical equi-probability.

Thus, *so far as the die is concerned*, equi-probability is assured by the balance of the parts. Let us now consider the motion of the hand that rolls the die. Can it be said that this motion is the proper effect of a proper cause, that it is determined in the sense in which a proper effect is determined, that is, according to a *de jure* and not only a *de facto* determinism, by the will of the gambler? Or should it be said that it is an effect of chance? A distinction is needed, a very simple one indeed, and one that can be easily verified in daily experience. Looked at from a distance, the action of rolling a die is a proper effect of the will of the gambler; on closer examination, it is the result of a multiplicity of nonunified proper causes. For the sake of simplicity, let us mentally suppress the dice cup and the particularities of the die as a missile. When I am about to throw a missile at a target, I form in my mind a motor scheme which I try to impart to my muscles. This scheme is the formal predetermination of the movement impressed upon my muscles and upon the missile, and insofar as this movement actually conforms to the representation meant to direct it, it is a proper effect and not a chance result. But as soon as a muscular movement is considered minutely, it becomes clear that the representation meant to direct it is not the only *form* that influences it. Moreover, the other forms involved in the concrete elicitation of a movement are never perfectly unified by the form that is meant to be directive. A concrete movement is a compound whose components are, *principally,* the force directed by the intentional scheme and, secondarily, forces directed by physical or psychological forms that always escape to some degree, no matter how small, from the directive intention. No matter how skillfully I control my movements, my motor scheme is never the only determining principle of my gesture; uncoordinated images, reflex stimulations, and imponderable resistances bring about slight discrepancies between the conceived scheme and the scheme that is actually realized. The most skilled motion is, in spite of all, a "faulty result" in the sense of Freud. It is a result brought about by nonunified forces. Training and skill may considerably reduce, but not eliminate, the part played by such parasitical forces.

Let us now remark that the nonunified system consisting of the intentional motor scheme and the parasitical forces is extremely unstable. Is it possible to conceive the same motor scheme twice? Apparently not, because the conception of the motor scheme is itself subjected to changing conditions. But even if it were possible, the disturbances brought about by collateral forces will never be entirely the same on two occasions, for the obvious reason that they depend upon ceaselessly changing circumstances—sensorial environment, position of the organs, etc. Thus, the impulsion imparted to the die, as it is repeated, assumes constantly different magnitudes and directions inasmuch as it results from an ensemble of forces each of which changes from instant to instant and whose relations to each other are extremely variable. We have assumed the most simple disposition; in fact, we complicate things and we multiply the nonintentional forces by shaking the die in the cup;

moreover, we weaken the motor scheme by acting absent-mindedly. Continual change in impulsions brings up now one side and now another. No preference for a special frequency of a particular side is contained in the ensemble of the causes, which repeats itself without ever being exactly the same twice. Thus physical equi-probability, which by hypothesis was already realized on the part of the die—by the regularity of the latter—is also realized on the part of the rolling, though not with an absolute necessity. Even though no preference for any particular face is contained in the successive series of the causal dispositions, a particular side may, by accident, come up more frequently. But by reason of the relative rarity of the accident, the relative importance of this frequency, which is truly abnormal because it does not correspond to any essential factor, will decrease with the number of trials. Thus we understand why there is a discrepancy between *a priori* and *a posteriori* probability and why this discrepancy increases in absolute value and decreases in relative value as the series of trials goes on.

Let us sum up the results of this analysis: At the point of departure we have a factor of order, the regular six-sided die. Through its cubic shape, it implies a definite law of mathematical probability,

$$p = \frac{1}{6}$$

[(1), *order*], and through its regularity it establishes, insofar as it is its own concern, the physical equi-probability [(2), *order*] which will allow, under certain conditions, the triumph of the law of mathematical probability. On the part of the gambler we have: (a) the will of the gambler as ruled by a motor scheme which is definite although it may be but weakly and partially conscious [(3), *order*]; (b); (c); (d) . . . , the parasitical stimulations of which we have spoken. Each of them is ruled by a definite form [(4), *order*]; the joint action of these forces brings about a chance effect which is determinate in fact but not *de jure* [(5), *disorder*]. The gesture is repeated but never in identical fashion; the series of its repetitions is altogether disorderly by reason of the variability of its causes and of their joint action [(6), *disorder*]; this very disorder guarantees that no side of the die enjoys any privilege, thus, physical equi-probability is realized [(7), *order*]. But this physical equi-probability is guaranteed only in *principle* by the lack of any factor that would privilege one of the possible results; in fact, there will be privileged results and a discrepancy between *a priori* and *a posteriori* probability [(8), *disorder*]. But this discrepancy has no proper cause and, consequently, is subjected to the law of relative rarity which governs chance events; it decreases in relative value as the series of the trials increases [(9), *order*].

This description suffices to show that disorder is not primitive in any way and that the final regularity of the statistical result is intelligible only by virtue of the principles of order primitively posited. The proximate, but always increasing, coincidence of *a posteriori* and *a priori* probability would not be obtained if the die were not regular; moreover, the die would not be rolled at all, whether regularly or irregularly, and nothing would happen if the will of the gambler were not determined by a motor scheme. The realization of the motor scheme would not be modified by parasitical movements if the latter were not stirred by stimulations each of which has its own determination. The phase of disorder, namely, irregularity in the movements of rolling, is the instrument of order, not its proper cause; by assuring physical equiprobability the phase of disorder releases the mathematical order embodied in the shape of the die,

$$p = \frac{1}{6}.$$

Here, as well as in a causal system, we start from order and arrive at order, but the process is complicated by a phase of disorder whose role is instrumental.[48]

INDETERMINACY IN PHYSICS

Before we consider indeterminacy in physics and its philosophic signification, it seems necessary to describe the diverse standpoints that a philosopher can adopt in a study of determinism.

The philosophy of determinism primarily pertains to the discipline called philosophy of nature or philosophical physics. The object of this discipline being nature itself, we ask, among other questions, whether the course of natural events is necessary or contingent, whether it involves both necessity and contingency, and if it does, in what sense it is necessary and in what sense contingent? We attempt to answer these questions by the theories of change, causality, and chance. In this philosophical interpretation of the physical world, all regulating concepts are suffused with an ontological meaning. Necessity and contingency and each kind of cause are defined in relation to being—more properly, to mobile being—the proper object of the philosophy of nature. No wonder that the positive scientist feels puzzled by such expositions and needs an unusual amount of good will to follow them. He reads such words as "cause," "law," and "determination," and he does not recognize the meaning that they customarily convey to him. Sometimes he even runs into such frightening words as "being" and "essence" and feels strongly tempted to utter the adjective that the Vienna Circle has so abused: *meaningless*. Let him control his impatience, however, and give credit for a while to philosophers. At least some of them hold resolutely that along with their ontological, or philosophical, science of nature, a nonontological and, in a way, anti-ontological physics is possible and necessary. These philosophers are convinced of the universal validity of such notions as necessity and causality. Yet they reject the silly theory that these notions could retain in a nonontological science the meaning that they have in a science in which all is defined in relation to the being of things. When the scientist declares that he does not recognize in the usage of the philosophers the meaning that he is used to giving words, he merely illustrates the so often voiced truth that the regulating concepts of all knowledge are primarily conceived—by common sense and by the philosophers—in relation to the being of things, and must undergo definite transformations before they can operate in a nonontological system.

The expression "philosophy of determinism" may also designate a chapter of the critique of knowledge. Here the object considered directly is no longer nature in its necessity or contingency, but the concept of determinism and the ways in which we use it in order to express the real in our diverse conceptual systems. This critical point of view obviously cannot be reduced to a psychological point of view. Psychology studies the idea as a subjective state of the soul; the critique of knowledge studies the relation of the idea to reality, its truth value, and the kind of truth it involves. The critical point of view is predominant throughout the present study. Let us note, however, that the judgment bearing on the relation of the idea to reality depends on the cognition of the corresponding reality. I cannot pass a critical judgment on the idea of determinism and the way in which we employ it in our different conceptual systems if I am in ignorance about everything concerning the real determination or indeterminacy of things. That is why we have been induced to include certain developments pertaining to the philosophy of nature in this study.

But if the idea of determinism can be integrated in many conceptual systems, that of common sense, that of philosophy, that of positive science (and here subdivisions would be possible in relation to the different types and periods of positive knowledge), it is appropriate to spell out exactly what notion of determinism we mainly have in mind. The entire movement of this study testifies that our principal objective is to establish the meaning of determinist and indeterminist ideas in positive science. If in certain passages we have tried out a criticism of the use of the notion of necessity in philosophy, if we have tried to describe the way in which the philosopher thinks about determinism, it is because that seems indispensable to us in order to establish a clear criticism of the positive (I do not say positivist) notion of determinism. Now we propose to devote our complete attention to the study of the meaning invested in the idea of determinism in positive science and especially modern physics.

When a contemporary physicist speaks of indeterminism, or indeterminacy, the first question for philosophic criticism is: How does he conceive the determinism both in relation and in opposition to which he speaks of indeterminism? It is up to the physicists to answer, and their answer, at first, seems to be unanimous: a process is determinate or, as they also say, causally conditioned when and only when such and such ulterior phases can be predicted on the basis of a system of initial data. In a more technical language, the physicist speaks of a deterministic system whenever, by the very fact that we know the position and velocity of a material point at the initial instant $t1$, we are able to calculate its coordinates at the ulterior instant t. Thus, determinism is conceived as a possibility of certain and exact prediction.

If such is the definition of determinism, an absolute determination of scientific phenomena would require absolutely precise measurements as well as an absolute isolation of the factors under consideration. As Louis de Broglie writes:

> In practice, however, there invariably arises indeterminateness that is in a sense accidental, since it is due to the imperfections of our methods of measurement. Actually, the co-ordinates and the initial velocity of any moving body are never known with absolute precision; all that we can say is that they fall within certain limits, generally very narrow, which give the exactness with which these magnitudes have been measured. From this slight indeterminateness of the initial data, however, there follows an indeterminateness in our predictions of the positions and final velocities of the moving body, this indeterminateness generally increasing with the lapse of time. But I repeat that according to classical notions this indeterminateness is contingent, so that it ought to be eliminated completely if only we succeeded in progressively perfecting our methods of measurement.[49]

Let us insist upon this point: this indetermination, because of its merely accidental character, does not raise any question such as would involve principles. Physics entered its "crisis of indeterminism" when a new kind of indetermination was evidenced, a kind that must be termed essential, in relation, of course, to the definition of determinism as the possibility of certain and exact prediction. Thus de Broglie continues:

> Now the new concepts introduced by contemporary physicists are the following: Beginning with the idea that every observation necessarily introduces some degree of disturbance into the phenomenon under investigation, they conclude, on the basis of an acute analysis, that even if we possessed infinitely perfect measuring instruments, it would be impossible to ascertain simultaneously and with absolute exactness both the position and the velocity of a corpuscle; quite apart from the contingent indeterminateness already referred to, there would always be an essentially irremovable indeterminateness.[50]

In the same vein, let us quote Eddington:

> Let us take the simplest case in which we think we can predict the future. Suppose that we have a particle with known position and velocity at the present instant. Assuming that nothing interferes with it we can predict the position at a subsequent instant. (Strictly the non-interference would be a subject for another prediction, but to simplify matters we shall concede it). It is just this simple predic-

tion which the principle of indeterminacy expressly forbids. It states that we cannot know accurately both the velocity and [the] position of a particle at the present instant.

At first sight there seems to be an inconsistency. There is no limit to the accuracy with which we may know the position, provided that we do not want to know the velocity also. Very well; let us make a highly accurate determination of position now, and after waiting a moment make another highly accurate determination of position. Comparing the two accurate positions we compute the accurate velocity—and snap our fingers at the principle of indeterminacy. This velocity, however, is of no use for prediction, because in making the second accurate determination of position we have rough-handled the particle so much that it no longer has the velocity we calculated. *It is a purely retrospective velocity.* The velocity does not exist in the present tense but in the future perfect; it never exists, it never will exist, but a time may come when it *will have* existed.[51]

Here is a particularly clear exposition by Gustave Juvet:

> If one admits that the electron is a spherical particle whose dimensions are ten thousand times smaller than those of the atom of hydrogen—which seems to be fairly well established by experiments of shooting through atoms—there can be no question of seeing it, or of having at least a photographic picture of it, unless it is illuminated with very short wave light. Suppose that this electron is compelled to follow a straight line. Its position on this line cannot be observed with absolute accuracy. Indeed, the error which can be made, for instance, in a microscopic observation, with a lighting of definite wave lengths, is proportional to this wave length and inversely proportional to the sine of an angle which will be called the opening of the microscope, as we learn from physical optics.
>
> But the light used is made of photons. The electron will not be visible unless a photon hits it. Whereas one hoped to study the movement of the electron on its rectilinear trajectory, this electron undergoes an impact whose effects, which were thoroughly studied by Arthur H. Compton, are similar to those of a mechanical impact. This phenomenon, known as the Compton effect, is even a proof of the corpuscular nature of radiation. This impact is the stronger as the momentum of the photon is greater. The momentum of the electron will be altered in proportion to the received impulse which, as we know from experience and theory, is equal to the constant of Planck divided by the wave length of the light

used. But the opening of the instrument does not allow us to observe this alteration of speed with absolute accuracy. What can be said is that the light which illuminates the electron modifies its momentum in such a way that the possible error, concerning this momentum, is inversely proportional to the wave length and directly proportional to the sine of the opening.[52]

According to Herbert Dingle, the principle of indeterminacy

> gives something approaching a definite limit to the extent to which the ordinary concepts of "position" and "momentum" can be applied to ultimate particles such as the electron. An electron can be said quite definitely to have a position in space if we do not attempt to consider it as having momentum, and it can be said quite definitely to have a momentum, measured in the ordinary way in terms of mass, space, and time, if we do not attempt to consider it as having a position in space at any instant. If, however, we try to apply to it simultaneously the conceptions of momentum and instantaneous position in space, we are attempting the impossible; the electron is such that in relation to it the two concepts are mutually contradictory.[53]

In brief:

1. There is in microphysical observation a margin of indeterminacy, and it is not accidental.

2. The cause of this indeterminacy does not lie in the relation between the initial and the ulterior state, but in our inability to obtain simultaneously all essential data concerning the initial state.

3. This impossibility follows upon the disturbance necessarily brought about by the observer.

4. This disturbance is rendered inevitable by the corpuscular nature of light. One must illuminate what one wants to observe, but to illuminate an electron is to throw at it a corpuscle which causes it to deviate from its trajectory—the Compton effect.

What are the philosophic consequences of this state of affairs? This question can be considered from the standpoint of the philosophy of nature or from the standpoint of the philosophy of science. From the first standpoint, it becomes: Does microphysics contribute any discovery of such significance as to force a reform upon our view of physical necessity? This question seems to call for a negative answer. Philosophy asserts that the natural event, inasmuch as it is predetermined in its proper cause, enjoys a *de jure* necessity and can be foreseen in its cause. But such foreknowledge is subject to two reservations: we may not be able to penetrate the cause so completely as to perceive in it the predetermination of its proper effect, and we can never remove the contingency of disturbance with absolute certainty. But the chance event, which is not predetermined in any cause, which does not enjoy any *de jure* necessity, and which, consequently, cannot be the object of scientific inquiry—for science is cognition through proper causes—still can be predicted in an historical sense by reason of the *de facto* determinism to which it is subjected. Again, the possibility of predicting is qualified by two reservations: a chance event cannot be safely predicted unless the nonunified system of the initial data from which it results is exhaustively known; moreover, interference by a factor foreign to the system of the initial data can invalidate even a well-grounded prediction.[54] Both in the case of the chance event and in that of the natural event, the theoretical conditions of an absolutely certain prediction are never satisfied, and scientific prediction will always be affected by some uncertainty, which uncertainty can be indefinitely reduced by improved methods of investigation and a more complete isolation of the investigated systems.

The essential indeterminacy of which the new physics speaks seems to boil down to the fact that in some domains of natural inquiry observation necessarily disturbs the observed, so that the first condition of any rigorous prediction, namely, the accurate establishment of a system of initial data, is impossible. Thus we must give up applying to all natural things the *scientific* principle of causality and the scheme of scientific prediction. The recently revealed fact of indeterminacy concerns the relations between man and nature rather than the course of natural events. So, the important problem raised by the indeterministic ideas born of quantum mechanics does not pertain to the philosophy of nature; it pertains, rather, to the critique of scientific knowledge.

Philosophy Within Science

It is well known that Max Planck, whose discoveries on the discontinuous nature of light are at the origin of the new physical conceptions, never ceased to oppose the indeterministic interpretations of some of his colleagues. Not only is Max Planck a great physicist, but his studies on the problem of determinism evidence a profound philosophic sense.

In an article entitled "Causation and Free Will,"[55] Planck goes over the main philosophic theories of causality; he judges them insufficient and proposes to supplement the answer of the philosophers with the answer of science:

> Let us ask how does science, in each of its different branches, actually regard the problem of causation. Here it must be remembered that I am talking of specialized science as such and not of the philosophical or epistemological foundations on which it works. Does science as a matter of fact occupy itself exclusively with data immediately given by sensory impressions and their systematic organization according to laws of reason? Or does it at the very outset of its activities reach out beyond the knowledge given us by this immediate source and make, as it were, a jump into the metaphysical sphere?
>
> I do not think that there can be any doubt whatsoever as to the answer. The first alternative is ruled out and the second affirmed in the case of each special science.[56]

He clarifies the point at once:

> I have said that the first step which every specialized branch of science takes consists of a jump into the region of metaphysics. In taking this jump the scientist has confidence in the supporting quality of the ground whereon he lands, though no system of abstract reasoning could have previously assured him of that. In other words, the fundamental principles and indispensable postulates of every genuinely productive science are not based on pure logic but rather on the metaphysical hypothesis—which no rules of logic can refute—that there exists an outer world which is entirely independent of ourselves.[57]
>
> Of course there is the positivist theory that man is the measure of all things. And that theory is irrefutable in so far as nobody can object on logical grounds to the action of a person who measures all things with a human rule, and resolves the whole of creation ultimately into a complex of sensory perceptions. But there is another measure also, which is more important for certain problems and which is independent of the particular methods and nature of the measuring intellect. This measure is identical with the *thing* itself. Of course it is not an immediate datum of perception. But science sets out confidently on the endeavor finally to know the *thing* in itself, and even though we realize that this ideal goal can

never be completely reached, still we struggle on towards it untiringly. . . .

Having once assumed the existence of an independent external world, science concomitantly assumes the principle of causality as a concept entirely independent of sense-perception.[58]

Let us review the main points contained in his remarkable article:

1. There is no sharp separation between science and philosophy; there is a *philosophy within science.*

2. This philosophy is realistic and holds that a thing independent of the mind is the ultimate measure of the mind. Planck goes so far as to say that the ultimate goal of science is to know things *in themselves,* which seems to us equivocal and rash.

3. The principle of causality is independent of all sense observation.

It follows that the impossibility of expressing certain events in deterministic formulations does not weaken at all our belief in the universality of real determinations. This belief results not from any definite observation, but from a metaphysical requirement. On the nature of this requirement, Planck expresses himself rather confusedly; he may well attribute to it a certain degree of irrationality.

Under these conditions, what becomes of the definition of determinism as possibility of certain and exact prediction, which seems to be unanimously accepted by the physicists of our time? Planck does not disrupt the unanimity, but he soon declares that this definition of determinism has but provisional significance.[59] The observable and measurable determination that constitutes the possibility of certain and exact prediction is but the empirical effect and the sign of a deeper determination pertaining to the things themselves as independent of our minds. And if it is true that the first action of scientific thought is a jump into the metaphysical, it must be said, correspondingly, that this determination pertaining to the things themselves, this ontological determination, is the objective foundation of scientific certainty.

The meaning of the controversy between deterministic and indeterministic physicists is now quite clear. It does not seem to be exceedingly daring to assume that physicists are agreed on the facts materially considered. Their final divergences proceed from diversity in their *epistemological* conceptions, that is, from different ways of conceiving the relation of science to the being of things. Some of them consider that a proposition cannot have a scientific meaning unless it refers to observations and measurements that can be really effected according to definite rules of operation. These want an entirely disontologized science, a science exclusively related to the observable and the measurable, without any conceivable reference to the being of things. For them, to say that the empirical or, better, empiriometrical determination is but the criterion of a real determination is a proposition devoid of scientific meaning; they hold that a real determination which would be distinct from the empiriometrical determination does not exist for scientific thought. From a scientific point of view, a determination modified by such ontological reference is like a quantity multiplied by zero. Every scientifically meaningful determination is empiriometrical, and when empiriometrical determination is lacking, we can talk only of indeterminacy.

By contrast, for Max Planck and those of his spirit, the assertion of determinism retains its scientific significance in spite of the restrictions imposed, in the new ways of physical research, upon the possibility of predicting events with certainty and exactness. Science is not conceived by them as a system of entirely disontologized cognitions but rather as an empiriometry supported by an ontology. The problem of the philosophic meaning of physical indeterminism ultimately resolves into the question whether positive science ought not to free itself from every ontological implication and from every reference to being.

In the examination of this question, it would be fitting to use first the method illustrated by Meyerson in his study of the products of scientific thought, in which he seeks to disengage the rules spontaneously obeyed in the actual construction of positive knowledge. Much of the material for such an inquiry is found in the works of Meyerson; an investigation of it would bring forth two main facts. On the one hand, positive science pursues indefatigably the work of its own disontologization; it endeavors to be no longer a philosophy of nature. As we reread Newton, we are astonished to notice that definitely ontological statements loom large in this Newtonian science which was considered by generations as the accomplished type of positive knowledge. Our astonishment gives an idea of what a long way it is from Newton to our contemporaries. On the other hand, we notice that the thinkers who are the most determined to purify science of any ontological residue run into apparently insuperable resistance, and we understand that their undertaking, if it could be carried to the term expressly aimed at, would annihilate science and annihilate itself at the very instant of its triumph; the *absolute* rejection of a reference to being has the disadvantage of suppressing the possibility of speaking and thinking.[60]

One might have suspected that the case was such if one had paid better attention to the metaphysics of understanding. A second method, designed to supplement the preceding one, would start from the elementary analysis of intellectual operations and, through a process of successive differentiations, would lead us from the general critique of the understanding to the general critique of the particular ways of the scientific intellect in each particular science. Let it be said that the movement of positive science seems to be animated by antinomic tendencies, both of which are sound and whose conflict is an indispensable factor of progress. One of these tendencies aims at the ontological interpretation of nature and is common to science and to philosophy. The other pertains specifically to positive science; it aims at expressing every scientific object in terms of observations and measurements. The first tendency is constantly repressed by the second, and during the phases of smooth development it remains mostly unconscious. But as soon as a crisis occurs, physicists are caught philosophizing.[61]

But if it is true that there exists a philosophy within positive science, we must conclude that the philosophy of nature admits of two states. It exists in a state of disengagement, of clarity, and of consciousness in the discipline that bears this name and is the work of the philosopher. It exists in positive science obscurely and vitally. The philosophy of the philosopher receives from the philosophy of the scientist healthful stimulation. Conversely, the philosophy of the scientist cannot disengage itself from the state of obscurity in which it is kept by the pressure of positive requirements and become critically conscious of itself unless it becomes a philosopher's philosophy and submits to the general laws of philosophic disciplines.

Finally, if we consider, from a psychological standpoint, that which takes place in the subjectivity of the scientist and in the scientifically minded public, attention should be called to the great part played by a factor not reducible either to the positive or to the ontological component of positive science. This third factor pertains to imagination; it is nonrational, though in a constant relation of mutual influence with the rational processes which it accompanies and by which it is penetrated. We propose to call it the *cosmic image*.[62]

Every great scientific epoch is characterized, for the historian of civilization, by the predominance of a certain cosmic image. Thus, in biology, Linnaean science is characterized by the image of everlasting patterns, whereas the image which illustrates the Lamarckian and Darwinian epoch is that of patterns that change unnoticeably. The great error of many interpreters is to confuse the cosmic image by which science is accompanied in the subjectivity of the scientist with the philosophy immanent in science. Revolutions in the properly positive part of positive science always entail a change of cosmic image, but they do not necessarily entail corresponding revolutions in the philosophic foundations of science.[63] Thus, the changes which recently took place in the positive systematization of physical knowledge as a result of the impossibility of applying to microphenomena the scheme of certain and exact prediction altered radically the cosmic image that haunts the minds of the physicists. Yet it does not seem that it should modify in essential fashion our ontological vision of nature, in which the concepts of substance, causality, and finality are fundamental.

Notes

1. "We ought then to regard the present state of the universe as the effect of its anterior state and as the cause of the one which is to follow. Given for one instant an intelligence which could comprehend all the forces by which nature is animated and the respective situation of the beings who compose it—an intelligence sufficiently vast to submit these data to analysis—it would embrace in the same formula the movements of the greatest bodies of the universe and those of the lightest atom; for it, nothing would be uncertain and the future, as the past, would be present to its eyes." Pierre-Simon Laplace, *A Philosophical Essay on Probabilities* (New York: Dover, 1951), p. 4. As we shall see later on, this text so often quoted is not content just to assert that every event is

capable of a rigorous prediction; it also asserts that all of the parts of nature are governed by a unique immanent law.

2. Cf. *Degrees of Knowledge*, pp. 82 and 83–84.

3. For a further discussion of this distinction, see the appendix on *Habitus* in Yves R. Simon, *An Introduction to Metaphysics of Knowledge*, trans. Vukan Kuic and Richard J. Thompson (New York: Fordham University Press, 1990), pp. 159–63 [ed. note].

4. Antoine Augustin Cournot, *An Essay on the Foundation of Our Knowledge* (New York: Liberal Arts, 1956), pp. 40–41.

5. Ibid., p. 41.

6. Ibid., pp. 42–43. See also *Traité de l'enchaînement des idées fondamentales dans les sciences et dans l'histoire*, rev. ed. (Paris: Hachette, 1911), pp. 64ff. *Matérialisme, vitalisme, rationalisme*, rev. ed. (Paris: Hachette, 1923), p. 314. In this latter work Cournot makes the observation that in order to exclude chance from the universe, it is not enough to assume that nature obeys a small number of immutable laws; assume, rather, that there are only *two* laws of nature, and chance has its place in the world order, if these two laws are perfectly independent of each other. On the relation between Aristotle's theory of chance and Cournot's, consult Gaston Milhaud, "Le hasard chez Aristote et chez Cournot," *Revue de Métaphysique et de Morale*, 10 (1902), 667–81.

7. An objection formulated against Cournot's theory by the psychologist Henri Piéron usefully calls attention to what is, in my opinion, the principal merit of this theory. According to Piéron, Cournot's definition of chance violates the definitional rule *soli definito;* it includes all the facts concerning chance, but also a number of facts that no one attributes to chance. "Would I say that it is a chance occurrence if I found a boat while passing by a lake? However, there is in that case an encounter between independent series of phenomena. But should I at that precise moment have the keen desire to take a boat ride and I will declare that there is involved in that fact a lucky chance, something completely out of the ordinary. Would you attribute to chance your meeting a dog on your way, although this is an unforeseen encounter? Obviously not. But a bicyclist who, wanting to avoid hitting a car, meets at that moment a dog that causes him to fall will curse the unlucky chance that put a dog in his way. Cournot's definition, then, seems inadequate because it is too general, and it is too general because it is incomplete. In fact, one can say that what is lacking is the human factor which really seems essential, a subjective factor, a relation to our ego which places a limit on the objectivity of the phenomenon." "Essai sur le hasard: La psychologie d'un concept," *Revue de Métaphysique et de Morale*, 10 (1902), 688. In other words, the common notion of chance implies a relation to human beings which clearly shows that it remains a practical notion. It

is the philosopher's duty to divest concepts relating to natural processes of the connotations that the interest we take in our personal business adds to them.

8. For the sake of a clear exposition, and because it is impossible to examine all difficulties at the same time, we take the term "predictability" here in a broad sense as signifying the *possibility of prediction regularly confirmed by the event.* We shall not fail to inquire further along if the chance event is, properly speaking, predictable.

9. Maritain, *Degrees of Knowledge,* pp. 28–30 [note updated, ed.].

10. See Chapter 2, "Science and Systematic Knowledge," below, pp. 67–73 [ed. note].

11. *Commentary on Aristotle's On Interpretation (Peri Hermeneias),* trans. Jean T. Oesterle (Milwaukee: Marquette University Press, 1962), I, lect. 14 [ed. revised note].

12. *The Meditations of Marcus Aurelius Antoninus,* trans. A. S. L. Farquharson (Oxford: Oxford University Press, 1990), p. 57 [ed. note].

13. Thomas Aquinas, *Commentary on Aristotle's On Interpretation,* I. lect. 14, p. 114 [ed. note].

14. Aristotle, *Physics,* I, 2, 185B19ff.

15. *Identity and Reality,* trans. Kate Loewenberg (New York: Dover, 1962), pp. 231, 252–53, 317. In a chapter of *The Great Dialogue of Nature and Space,* ed. Gerard J. Dalcourt (Albany, N.Y.: Magi, 1970) entitled "How We Explain Nature," Simon has this to say of Meyerson's position: "Comte retained the word explanation but it is soon understood that the word no longer has any substance. Explanation through law alone, without cause, is not understood by anybody as an explanation. So if science does not trace consequences to principles, and effects to causes, what is it good for? We have seen what it is good for in the mind of Comte and in the mind of Mach. Meyerson, in *Identity and Reality,* holds that the spontaneous reality of the science of inquiry in the nineteenth century is constantly at variance with the positivists' understanding of science. For Comte forbad the scientists to indulge in what he called metaphysical curiosity, to look for causes, reasons, grounds and so on. But actually they do nothing else from dawn to dusk, and oftentimes till late at night; they are always looking for explanations, for causal relations, for reasons and for grounds. And what do they do in their effort to explain nature? *Identity and Reality* can be summed up in these words: The whole history of the scientific mind shows that to explain is to identify. *To explain is to identify.* If we are concerned with change, to explain will consist in showing that underneath appearances which involve novelty there is something which remains identical. If we are concern not with change but with sheer multiplicity, to explain will consist in showing that things are not so diverse as they look, that if

you scratch a little you will notice that diversity is superficial and that underneath a diversified surface there is a background of homogeneity." Pp. 27–28 [ed. note].

16. In Saint Thomas's words: "Besides, the large number and variety of causes stem from the order of divine providence and control. But, granted this variety of causes, one of them must at times run into another cause and be impeded, or assisted, by it in the production of its effect. Now, from the concurrence of two or more causes it is possible for some chance event to occur, and thus an unintended end come about due to this causal concurrence. For example, the discovery of a debtor, by a man who has gone to market to sell something, happens because the debtor also went to market. Therefore, it is not contrary to divine providence that there are some fortuitous and chance events among things." *On the Truth of the Catholic Faith: Summa contra Gentiles,* trans. Vernon J. Bourke (Garden City, N.Y.: Doubleday, 1956), III, 74, p. 247. "For when a cause is efficacious to act, the effect follows upon the cause, not only as to the thing done, but also to its manner of being done or of being. . . . Now God wills some things to be done necessarily, some contingently, that there might be order in things for the perfection of the universe. Therefore to some effects He has also attached necessary causes, that cannot fail; but to others detectible and contingent causes, from which arise contingent effects. Hence it is not because the proximate causes are contingent that the effects willed by God happen contingently, but because God has prepared contingent causes for them, it being His will that they should happen contingently." *Summa Theologiae,* I, 19, 8. "So far then as an effect escapes the order of a particular cause, it is said to be a chance occurrence or fortuitous in regard to that cause; but if we regard the universal cause, outside whose range no effect can happen, it is said to be foreseen. Thus, for instance, the meeting of two servants, in such a way that the one does not know about the other's task, although it appears to them a chance occurrence, has been fully foreseen by their master, who has purposely sent them to meet at the same place." Ibid., I, 22, 2 ad 1. "Things are said to be chance occurrences as regards some particular cause from the order of which they escape. But as to the order of divine providence, nothing in the world happens by chance." Ibid. I, 103, 7 ad 2. [Translations from the *Summa Theologiae* are by the author, in some instances slightly altered by the editors.] "Notice that both the fortuitous and what happens by chance are relative things. Consequently, an event that, considered in itself and with no further specification, is fortuitous is not necessarily fortuitous in relation to every agent. Between things absolute and things relative there is the following difference: when a thing is unqualifiedly such and such according to an absolute predicate, it never ceases to be such and

such, whatever may be the term to which you relate it. But when a thing is such and such, in itself and unqualifiedly, according to a relative predicate, it does not possess this predicate in relation to every conceivable term. On the contrary, it possesses this predicate only in relation to its correlative. It may even deserve an opposite relative predicate, in relation to another correlative. For example, white is unqualifiedly similar to another white, but, if compared with black, is dissimilar. Thus, the fact that things are called fortuitous or chance occurrences in relation to created nature and intention does not exclude their being described as non-fortuitous and properly intended when they are related to divine intention. However, inasmuch as everything is what it is by reason of its proper causes and must be interpreted in relation to its proper causes, those events, considered in themselves, are and are said to be by chance or fortuitous in an unqualified sense." Cajetan, *In Summa Theologiae*, I, 22, 2 ad 1. It is interesting to compare these Thomistic texts, so remarkable for their accuracy, with the confused approximations of Bossuet on the same subject: "Thus God reigns over every nation. Let us no longer speak of coincidence or fortune; or let us use these words only to cover our ignorance. What is coincidence to our uncertain foresight, is concerted design to a higher foresight, that is, to the eternal foresight which encompasses all causes and all effects in a single plan. Thus all things concur to the same end; and it is only because we fail to understand the whole design that we see coincidence or strangeness in particular events." *Discourse on Universal History*, trans. Elborg Forster (Chicago: The University of Chicago Press, 1976), p. 374. [A more literal translation would replace "coincidence" with "chance," for the word used is *le hasard*]. If these beautiful sentences were meant to say what they do say, we should understand (1) that the reality of chance is incompatible with providential government, and (2) that in order to have chance displaced by intention, it suffices to substitute the consideration of the *whole* for that of the parts, which suggests the Stoic idea of a universal necessity immanent in the whole. Again, chance disappears only on the level of the first cause, which is *transcendent* to the whole universe.

17. See the meaning of this term in *The Material Logic of John of St. Thomas*, trans. Yves R. Simon, John J. Glanville, and G. Donald Hollenhorst (Chicago: The University of Chicago Press, 1955; repr. 1965), pp. 78–79 [ed. note].

18. Aristotle, *Physics* I, 8.

19. "In Aristotle, it is above all from something that is in potency that the effective and real body is derived. It is derived from it; that is, according to the Stagirite's logical system, it must be deduced from it, which in turn can mean, we have seen, only one thing, if we abandon the metaphysical formalism and if we seek how that might be translated

by properly physical theories, namely, that there is a basic identity between the two states." Émile Meyerson, *De l'explication dans les sciences,* 2nd ed. (Paris: Payot, 1927), p. 320. There we find a rather clear formulation of the radical misunderstanding to which the concept of potency will always give rise. Because potency is unintelligible in itself and is intelligible only in relation to act, we will always be tempted to reduce potency to a kind of act, to an act *enclosed,* "basically identical" to what it will be when a process more apparent than real will have *unfolded.* In order to safeguard the genuine concept of potency, one should be aware of this danger of perversion and agree to an effort at restoration which will always have to be begun anew. In this interminable struggle it is good to keep in mind some particularly clear examples that will play the role of points of reference. Let us imagine a newly born infant and a newly born monkey. We say that the infant is potentially a geometrician and that the monkey is not. The difference is a real one. Yet the infant is as completely ignorant of geometry as the monkey is. There is not the slightest rudiment of geometrical knowledge in its mind. There is a strong temptation to conceive the potentiality of being a geometrician as an actual preformation of the science in the mind. Therefore let us understand that from the viewpoint of geometrical knowledge the newly born infant and the little monkey are strictly equal. It is impossible to speak of a "basic identity" between the state of potentiality and the state of actuality.

20. See Yves R. Simon, "On Order in Analogical Sets," *The New Scholasticism,* 34, No. 1 (January 1960), 1–42 [ed. note].

21. The complete expression of the principle of efficient causality is found in three axiomatic formulations that represent three successive moments in the mind's itinerary: *quidquid movetur ab alio movetur; omne agens agit secundum quod est actu; omne agens agit sibi simile.* The first formulation asserts the need to recur to an *other* in order to understand the *same.* The second formulation indicates that the *other* that we summon owes its privilege only to the actual possession of what the *same* has by itself, only in potency. The third formulation, by pointing out the specific unity existing between the *other* and the *same* (considered as a support for a new determination) restores unity amid diversity and shows the explanatory value of the causal relationship. See Yves R. Simon, *Freedom of Choice,* ed. Peter Wolff (New York: Fordham University Press, 1969; repr. 1987, 1992), pp. 127–35 [ed. note].

22. "The efficient cause and the form of the effect are the same in species inasmuch as every agent effects something similar to itself; but they are not numerically the same, because the maker and the thing made cannot be identical." Saint Thomas, *Truth* III, trans. Robert W. Schmidt (Chicago: Regnery, 1954), p. 19. "Now, since every action

proceeds from an agent with a certain similarity to the agent, as hot things heat. . . ." Ibid., p. 165. ". . . and hence the proximate end that is proportioned to an agent falls into the same species as the agent both among natural things and [among] artificial things. For the form of the art through which the artisan works is the species of the form that is in matter, which is the end of the artisan." Saint Thomas, *On the Truth of the Catholic Faith: Summa contra Gentiles,* trans. Anton C. Pegis (Garden City, N.Y.: Doubleday, 1955), I, 72, p. 242. "We see that every agent insofar as it is actual and perfected produces something similar to itself." *Summa Theologiae,* I, 19, 2. "Now, everything that produces something by its action produces something similar to itself, as to the form by which it acts." Ibid. I, 41, 5. This would be an appropriate place to point out a strange error made by Meyerson. After having revealed the idea at the beginning of *Identity and Reality* that will dominate his whole work, that is, that the causal relation is reduced to a relation of identity, more precisely to a relation of temporal identity, he states that this way of thinking about causality does not apply to acts that emanate from free choice. "When, by an act of volition, I produce an external change, or when the believer attributes a phenomenon to the intervention of God (we have previously shown that these are associated concepts), it is certain that one does not hesitate to speak of cause and effect. But here no identity is possible, and, what is more, I have the immediate intuition of this. Not for a single moment can I nourish the illusion that my will is something analogous to the movement which it produces; here, then, is a concept of causality fundamentally different from that which we have just stated and which is based on identity. In order to mark the distinction we shall designate the latter concept as that of scientific causality, and we shall apply to the first the term theological causality, since, as we have just seen, the supposition of God's intervention in natural events makes use of it." *Identity and Causality,* pp. 46–47. It is clear that Meyerson misconstrues the emphatic precision of Saint Thomas: "it produces something similar to itself *as to the form by which it acts.*" Voluntary movement resembles the will insofar as it is determined, in the order of formal causality, by the practical judgment that declares the execution of such and such a movement as desirable. The principle of proper causality really applies as much to the world of freedom as it does to the world of nature. See Simon, *Freedom of Choice,* pp. 138–43 [ed. note].

23. These two predeterminations, all too often confused, are analyzed with remarkable clarity by John of St. Thomas. Cf. *Cursus Philosophicus: Philosophia Naturalis,* I, q. 13, a.1., ed. Beatus Reiser (Turin: Marietti, 1933), Vol. 2, p. 273 b, 16.

24. See Simon, *Introduction to Metaphysics of Knowledge,* pp. 1–4 [ed. note].

25. See Yves R. Simon, *The Tradition of Natural Law: A Philosopher's Reflections,* ed. Vukan Kuic (New York: Fordham University Press, 1965, repr. 1967; rev. ed., 1992), pp. 144–45 [ed. note].

26. Otis E. Fellows and Stephen F. Milliken, *Buffon* (New York: Twayne, 1972), p. 35 [ed. note].

27. "The philosophy of nature can be defined as a physical consideration whose conceptual instruments call for an ascending analysis, positive science as a physical consideration whose conceptual instruments call for a descending analysis. The very opposition of the two analyses provides an invaluable rule for the determination of the point of view prevailing in our studies about nature. Let us think of the ambiguous literature which stands on the borderline between philosophy and positive science. When a philosopher informed of positive science or a scientist interested in philosophy considers philosophical problems raised by the study of positive questions, the philosophical and the positive point of view appear successively in his expositions; generally the writer is not aware of the shift. The resulting confusion can easily be removed provided we carry out the analysis of a few key concepts. According as this analysis goes up or down, according as the concept demands to be explained in more and more characteristically ontological terms or in terms which refer more and more directly to definite experiences, we know whether we have to do with a philosophical or positive treatment." Yves R. Simon, "Maritain's Philosophy of the Sciences," in Jacques Maritain, *Philosophy of Nature* (New York: Philosophical Library, 1951), pp. 166–67. This classic article of Simon's has been reprinted and translated many times; for an updated English version see: *The Philosophy of Physics,* ed. Vincent E. Smith (Jamaica, N.Y.: St. John's University Press, 1961), pp. 25–39 [ed. note].

28. These investigations require a daring imagination. In Freud imagination often goes wild, and in his disciples goes even wilder. Simon discusses causality in Freudian psychology in *Freedom of Choice,* pp. 38–44 [ed. note].

29. The gentleman intended to say that certain facts "came . . . to light" and so he should have used the word *Vorschein*. Instead he used the meaningless word *Vorschwein*. The upshot of the slip is to connect the facts and piggishness, that is, something disgusting. Cf. Sigmund Freud, *The Psychopathology of Everyday Life* (London: Hogarth, 1960), p. 57*n*1 [note updated, ed.].

30. Raymond Aron, *Main Currents in Sociological Thought,* trans. Richard Howard and Helen Weaver, 2 vols. (London: Weidenfeld & Nicol-

son, 1965, 1967), II, 183. The phrase appears in a chapter devoted to Max Weber [note updated, ed.].

31. "Therefore we can say that light is a cause in this case; we certainly cannot say that it is an explanation." Edmond Goblot, *Essai sur la classification des sciences* (Paris: Alcan, 1898), p. 42.

32. "Hypotheses present us with a somewhat paradoxical situation. It is increasingly obvious that they are completely artificial and that the notion of atoms only explains a phenomenon on the condition that one has attributed appropriate properties to them in the first place to render this explanation. And yet the history of science teaches us that our trust in the certainty of laws depends on the nature of the hypothesis that we form to represent them. Scientists have a more or less definite feeling about this singular situation, and it is for that reason that many of them find it hard to resign themselves to the indifference or contempt that many geometricians display for hypotheses, for the scientists understand that physical laws would be missing something by losing their alleged physical causes." Georges Sorel, *Les préoccupations métaphysiques des physiciens modernes* (Paris: Cahiers de la Quinzaine, 1907), p. 43.

33. "From what has been said before, it can be concluded that demonstration, and science is the habitus of demonstrated conclusions, is always concerned with the incorruptible and the sempiternal. For where it is correct to predicate universally, it is necessary that the subject be incorruptible and sempiternal. As was said above, to predicate universally is not to say that something belongs or does not belong, but that something always belongs. Corruptible things, however, do not always exist. Therefore it is not correct to predicate universally in regard to them, but only about sempiternal things. But in the strongest demonstration in which both the premises are universal, it is correct to predicate universally of all. Therefore demonstration concerns the incorruptible and the sempiternal." *Summa totius logicae Aristotelis* in Saint Thomas Aquinas, *Opuscula omnia, genuina quidem necnon spuria melioris notae debito ordine collecta cura et studio*, ed. Pierre Mandonnet, 5 vols. (Paris: Lethielleux, 1927), V, IX, 7, p. 150. Although John of St. Thomas thought this was a Thomistic opusculum, an editorial note in *Material Logic of John of St. Thomas*, states: "Let it be recalled that the *Summa of the Whole Logic of Aristotle* is definitely not the work of St. Thomas." P. 610n26 [ed. note].

34. See also the discussion of virtual distinction in *Material Logic of John of St. Thomas*, pp. 78–79 [ed. note].

35. We offer this remarkable passage by Roland Dalbiez on the central role of the concept of tendency in any science of physical reality or metaphysics. "Strict positivism tells us that there are only events, and that 'tendencies' must be dismissed as 'mythological entities.' We need no other proof to show that this purely phenomenalist point of view is

untenable than the evidence of that unimpeachable positivist, Stuart Mill, who declares that all physical causes may be counteracted, i.e., that from a purely factual standpoint, no universal laws can be formulated. The philosopher who knows no tongue but that of strict phenomentalism is precluded from stating a single truly general law. Let us take the case of gravitation. If we are to invoke no 'mythological entity,' we must say: all weighty bodies fall. But this is completely untrue; there are any number of weighty bodies which do not fall because they are prevented. We must therefore either dismiss all general formulae, or say: weighty bodies *tend* to fall. But since no one has ever had, and never will have, sensory perception of a 'tendency', it is clear that our assertion 'weighty bodies *tend* to fall' is metaphysical. We might give any number of examples of this kind." *Psychoanalytical Method and the Doctrine of Freud,* trans. T. F. Lindsay, 2 vols. (London and New York: Longmans, Green, 1941; repr. Freeport, N.Y.: Books for Libraries Press, 1972), II, 22. Mill says: "These facts are correctly indicated by the expression *tendency.* All laws of causation, in consequence of their liability to be counteracted, require to be stated in words affirmative of tendencies only, and not of actual results." See *A System of Logic: Ratiocinative and Inductive* (Toronto: University of Toronto Press, 1973), Bk. III, Ch. X, p. 445. To which Dalbiez responds: "Perfectly true, but how, after writing this highly Aristotelian passage, could Stuart Mill still think himself a positivist?" *Psychoanalytical Method,* II, 22.

36. *The Value of Science* (New York: Dover, 1958), pp. 110–11. The original version is entitled *La valeur de la science* (Paris: Flammarion, 1904) [note updated, ed.].

37. The meaning of these expressions seems rather successfully specified by General Vouillemin in his blunt nominalistic formulations. "We would say that B is *causally* connected to A if B succeeds A without fail. Water begins to boil (B) every time one brings it to a certain temperature under a certain pressure (A). A large number of positive experimental outcomes have been carried out and up to now no negative experimental outcomes have been recorded. . . . We would say that B is *statistically* connected to A if B itself is not reproduced at a designated point in the course of the reproductions of A, but occurs in a constant proportion in the *groups* of repetitions of A. Only the proportion matters here." "Hasard et régularité," *Revue d'Artillerie,* 1937.

38. Throughout this translation Boyle's name will replace that of Mariotte. Mariotte's law is the same as Boyle's law. It is a law of thermodynamics which states the principle that, for relatively low pressures, the pressure of an ideal gas at constant temperature varies inversely with the volume of the gas [ed. note].

39. ". . . there is not and there cannot be a philosophy of nature in a

system like Plato's. On the one hand, you have *doxa,* opinion, which is concerned with the sensible world and its becoming; on the other hand, you have the world of eternal archetypes, the object of metaphysics. On one side you have opinion about the world of becoming and on the other, as science, you have mathematics and metaphysics: no scientific knowledge of nature, no scientific knowledge of the world of movement and time. Wherefore, when the philosopher tries despite everything to give an interpretation of this world and to rise above common opinion, he can proceed only with the help of myths. The use of myths to interpret sensible nature is really indispensable in Plato's philosophy. I think it can be generally said that every attempt to explain natural phenomena by the use of mathematical knowledge alone necessitates the recourse to explanatory myths." Maritain, *Philosophy of Nature,* p. 7 [note updated, ed.]. ". . . while sense perception grasps only the perpetually varying accidents, appearances which are here today and which will have disappeared tomorrow, Geometry knows permanent and eternal objects, the realities; in the study of these immutable objects, it prepares our souls for the use of intuition which alone can contemplate essences and, in particular, the pre-eminent essence, the supreme Good. This Platonic doctrine is characterized by the mistrust that it professes regarding the data of sense perception. . . . the only science worthy of this name is where the things known are immutable and eternal, as the truths of Geometry are, and as the ideas that intuition contemplates are." Pierre Duhem, *Le système du monde: Histoire des doctrines cosmologiques de Platon à Copernic,* 10 vols. (Paris: Hermann, 1913), I, 130–31. "If there is only science of the really real, and if all reality is immutable, there is only science of immutable things. Mathematics, which studies the unvarying properties of numbers and figures, and Theology, the intuitive contemplation of eternal ideas are the only two sciences that a Platonist can recognize. Aristotle, by a bold reversal of the most essential Platonic dogmas, defines a third science, Physics." Ibid., 135. "When the mathematician saves the appearances with the aid of a theory, Plato thinks that he applies to the appearance something of the certainty of which the Geometric method is capable; Aristotle believes, on the contrary, that the mathematical theorist elevates to the level of theory a part of the truth that the senses have grasped directly." Ibid., 140.

40. *Value of Science,* p. 87.
41. Ibid.
42. "The Law of Chance," in *Science and the Human Temperament,* trans. James Murphy and W. H. Johnston (New York: Norton, 1935), pp. 39–51.
43. Ibid., pp. 41–43.
44. Ibid., p. 44.

45. Ibid., p. 49.

46. Ibid., pp. 50–51.

47. *La probabilité dans les différentes branches de la science* (Paris: Hermann, 1937), p. 6.

48. "It has been maintained that the certainty which pertains to a whole presupposes a kind of determinism on the side of the components of that whole. To say the least, this is a strange assertion. After all we cannot deny successful predictions. I don't know which side of a die will turn up in individual throws. But this kind of ignorance does not stop me from predicting that out of six thousand throws, the six sides will be quite equally distributed. If this prediction is not verified, I will be suspicious about the die and not about the calculation. The only thing I have to know beforehand is the number of alternatives, the equal probability of these alternatives, the independence of individual throws and their number. If the alternatives are not equally probable, that is, if there is not a perfect indetermination within the bounds of the determined number of alternatives, my prediction will be false. Macroscopic determinism is conditioned by microscopic indeterminism. However, the kind of determinism at issue cannot be absolute, for that would assume an actually infinite number of components which is to say a contradiction." Charles De Koninck, "Réflexions sur le problème de l'indéterminisme," *Revue Thomiste*, 43 (1937), 400–401. If I understand de Koninck's account correctly there would then be only two phases to consider: the macroscopic phase of determinism and the microscopic phase of indeterminism. On the contrary, we would ask that one go back to a third phase, which could be called ultramicroscopic, that of initial determination. Is there an irreconcilable opposition between the views of De Koninck and our own, or is it just a matter of differences in the mode of exposition? No doubt it is too early to tell. In the introduction of the article quoted, De Koninck invited the reader to look for no more "than 'approaches' in his article—very scattered and very unspecified—which could contribute to a thorough systematic study of the problem." Ibid., 228. There is every reason to think that this phrase not only expresses the author's modesty, but also the judgment he passes on the real character of his work which is remarkable from a number of points of view. For my part, I would entreat the reader to consider that the present piece of work is also only a collection of efforts at an approach.

49. *Matter and Light: The New Physics,* trans. W. H. Johnston (New York: Dover, 1946), pp. 227–28 [note updated, ed.].

50. Ibid., p. 228.

51. Sir Arthur S. Eddington, *The Nature of the Physical World* (London: Dent, 1947), pp. 294–95 [note updated, ed.].

52. *La structure des nouvelles théories physiques* (Paris: Alcan, 1933), p. 135.

53. *Science and Human Experience*, p. 62. See also the lucid account of Arthur H. Compton, *The Freedom of Man* (New Haven, Conn.: Yale University Press, 1935), pp. 32–39.

54. Jacques Maritain in "Reflections on Necessity and Contingency," in *Essays in Thomism*, ed. Robert E. Brennan (New York: Sheed & Ward, 1942; repr. Freeport, N.Y.: Books for Libraries, 1972), pp. 27–37, sketches a theory of the prediction of the contingent. By "future contingents" he understands three kinds of event: (*a*) "imperfectly assured" natural events, that is, natural events that, differing in this from the course of the stars, are subject to the counteraction of factually existing factors; (*b*) chance events; and (*c*) free events. "Future contingents so defined are *not capable in themselves of being foreseen with certainty*. This is so either because the proper cause that is preordained to produce them is preordained to them only in an insufficient or variable or indeterminate manner, or because such a cause does not exist for them, or because they depend on a cause which, being exceedingly rich in causality, is master of its own determination. Now, to *foresee* or foreknow is to see or recognize a thing *in its cause*. Therefore, when the cause is of such a kind that it fails to make known with certainty the thing to which it is preordained, such a thing is not in itself capable of being foreseen with certainty. This, however, does not prevent future contingent events (natural events imperfectly assured in their causes, or again chance events), which are to be produced in the course of time, from appearing necessitated in fact with regard to the transfinite multitude of factors that we would have before us, were we able to take into consideration, as we said above, all the placement of causes of all the agents of the universe and their complete history (excluding, by hypothesis, the intervention of free agents). But, on the other hand, this multitude of factors is beyond the comprehension of any created mind, so that with regard to a finite intellect the future contingents of which we are speaking are not in themselves foreseeable with certainty. And on the other hand—and especially—to follow beforehand the course of all the historical events that will intersect in time by a detailed calculation (if that were possible) would not be to foresee or foreknow a thing *in its cause*. It would be to mentally live through the entire sequence of effects that antedate the thing, to exhaust the multitudinous array of events that led up to it; but it would not be to know the thing in the more simple light and, as it were, intelligible abridgement which is the cause or reason of the thing's being. In this sense it must be said in a *universal way* that the future contingents that we are discussing—although necessitated in fact with regard to the almost infinite number of actual factors in the universe

since the first moment of the existence of the universe—are not capable by themselves of being foreseen with certainty" (pp. 34–35). Maritain subsequently observes that a future contingent event of a physical type (a natural event or chance occurrence) can be predicted with certainty by accident. "Thus it may so happen, in certain relatively simple cases, that the mind embraces in the unity of its knowledge the plurality of independent causal series from which the occurrence of such a future contingent results. The meeting of two balls rolling down an incline is a chance event (dependent upon an irreducible pluralism in causality); and their meeting can be calculated with certainty if the conditions of the problem are sufficiently determined" (p. 35). As to free events, "it is for the best of reasons that they are incapable of being foreseen. Neither in themselves nor by accident can they be known beforehand with certainty, their nature being such that they are absolutely inamenable to certain foreknowledge, since they depend on no necessity, either *de jure* or *de facto*" (p. 36). I am in complete agreement with Maritain's views [in this regard, Simon's essay "On the Foreseeability of Free Acts," *The New Scholasticism,* 22, No. 2 (October 1948), 357–70, should be examined—ed. note] and would like to add the following remarks to them:

(*a*) It seems that the predictability of a natural event and that of a chance event should be treated in different ways. If the term "foresight" is taken in its proper sense, in other words, if foresight means foreknowledge, "to have advance knowledge of a thing in its cause," then the chance event is in no way predictable, because there exists no cause in which the chance event would be preformed and could be *seen* in advance. We can speak of foreseeing a chance event only in an improper and entirely historical sense, not at all in an essential sense. As to the natural event, as unsteady as one supposes it to be, it is preformed in its cause and can be seen in advance in its cause insofar as it is preformed in it. "Concerning future contingents, it should be said that considered in themselves, there cannot be scientific knowledge of them, but on the conditions that they are considered in their cause, there can be scientific knowledge about them, according as some sciences recognize the causes as being *certain inclinations to specific effects.*" Saint Thomas Aquinas, *Commentary in De memoria et reminiscentia,* lect. 1. Therefore it is possible to speak of prediction in regard to a natural event in an essential sense. This prediction is conditional, even highly conditional, if it concerns a complicated conditioned event. As to the very conditions without which the event that nature tends to produce will not be realized, their concurrence constitutes the object of foresight—or, to put it in a better way, of a prediction—of a completely historical kind. All and all, we have a prediction of a complex character, in which a core of prediction in an essential sense is accompanied by a coating of prediction in an historical sense. In the case of a chance event the core of essential prediction is absent. From a practical viewpoint the difference might be negligible; from a theoretical viewpoint there is a considerable difference.

(b) In regard to the proposition "a future contingent can be foreseen with certainty by accident," it seems indispensable to me to observe that the certainty under discussion remains conditional. It is, putting things in the best light, a certainty on the physical scale; it is not absolute certainty. That astronomical events are capable of being calculated, our experience verifies with impressive regularity. But is it possible to affirm that an eclipse duly calculated by astronomers is the object of an absolutely certain prediction? We forget that it concerns a contingent event because its human interest does not weigh heavily. But imagine an astronomical event of a kind that shapes our destiny. We would not hesitate to remember that it is contingent and can be impeded, at least by a miraculous intervention of the First Cause. Think of the nightmares of popular astronomy, the collision of the earth and a heavenly body, the wiping out of our planet by the tail of a comet loaded with noxious gases. If it happened that such a catastrophe was announced by scientists armed with calculations, we would not hesitate to pray that we would be spared this catastrophe, knowing full well that it is no more difficult for God to divert an asteroid from its trajectory than to preserve the existence of a gnat. In a passage in which he points out the uncertainties of popular consciousness in matters of determinism, Henri Poincaré asks: "Why is it that showers and even storms seem to come by chance, so that many people think it quite natural to pray for rain or fine weather, though they would consider it ridiculous to ask for an eclipse by prayer?" *Science and Method,* trans. Francis Maitland (New York: Dover, 1952), p. 68. His idea is that meteorological facts are just as determined as astronomical facts, and that if we refuse to bother God in regard to the latter it would not be reasonable to bother him more in regard to the former. That is a rather superficial interpretation of the metaphysical and religious state of mind. There are two things to be said in response to the question why one prays to obtain rain rather than to obtain an eclipse. First of all, the succession of a rainy season to a dry season does not have or at least does not necessarily have a miraculous character. Unless the believer demands a sudden downpour from a clear sky, when demanding rain, the believer does not seek for an event which is extraordinary *de jure*. In the same way, to pray for the healing of someone who is sick is not to ask for a miracle if the sick person has some chance of healing naturally. On the contrary, to pray in order to obtain an eclipse that the normal development of astronomical trajectories makes impossible would be to ask for a miracle. Secondly, it happens that we pray in order to obtain miracles, and this prayer is sometimes answered. But we know that it would be unreasonable to ask God for the favor of an extraordinary intervention when important human interests are not at stake. Also it does not occur to us to ask for a miracle for an object as futile as the pleasure of observing an eclipse. Moreover, we do not ask to be miraculously cured of a light cold. But if, instead of a cold, it is a matter of tubercular meningitis, and if, instead of an eclipse, it is a matter of an astronomic encounter capable of annihilating humanity, then we will not hesitate to seek our salvation in a miraculous intervention, thus bearing witness to the radical contingency of any created existential situation.

55. In *Where Is Science Going?* trans. James Murphy (New York: AMS Press, 1977), pp. 107–40 [note updated, ed.].

56. Ibid., pp. 136–37.

57. Ibid., p. 138.

58. Ibid., pp. 139–40. See also Yves R. Simon, "An Essay on Sensation," in *Philosophy of Knowledge: Selected Readings,* ed. Roland Houde and Joseph P. Mullally (Chicago: Lippincott, 1960), pp. 80–83, on what is and what is not an immediate datum of perception [ed. note].

59. "Causality in Nature," *The Philosophy of Physics,* trans. W. H. Johnston (New York: Norton, 1936), pp. 43–83.

60. Philipp Frank in a discussion of Planck's "ontological" conception, *Between Physics and Philosophy* (Cambridge: Harvard University Press, 1941), p. 135, examines the psychological argument that emphasizes that the desire to know the real world is the great stimulus to scientific research. He recalls that Friedrich Schiller always had a provision of rotten apples in his desk drawer because he found a stimulus to poetic creation in their odor. The reasoning is that the rotten apples are not incorporated by the poet into his worldview and there is no reason that the "real" be incorporated into the worldview constructed by the physicist. Let us be content to remark that the causal relation between rotten apples and poetic creation is obviously accidental. As to the relation between the real world and the mind's movement in its search for scientific truth, it is not evident that it can be reduced to a psychological accident. After having written that "the property *Fer* records the discovery of a connected whole of determined qualities and, above all, the discovery that the said property is regularly maintained, General C. E. Vouillemin, an exceptionally radical positivist, remarks in a note that "We have been imitating the ostrich in vain. All these expressions—a connected whole, complex, and so forth—hardly conceal the uneasiness before the horrible 'thing-in-itself'. The only thing that can be said is that science doesn't deal with it, and that by not dealing with it, it leaves

the patient to its uneasiness, which doesn't amount to dispelling it." "Sur la theorie physique," *Revue d'Artillerie* (March 1932), 224.

61. Albert Einstein, "Physics and Reality," *Ideas and Opinions* (New York: Crown Publishers, 1954), pp. 290–323 [note updated, ed.].

62. We find a very suggestive description of what we call the cosmic image in the following text of Alfred North Whitehead's: "Every philosophy is tinged with the colouring of some secret imaginative background, which never emerges explicitly into its trains of reasoning. The Greek view of nature, at least that cosmology transmitted from them to later ages, was essentially dramatic. It is not necessarily wrong for this reason; but it was overwhelmingly dramatic. It thus conceived nature as articulated in the way of a work of dramatic art, for the exemplification of general ideas converging to an end. Nature was differentiated so as to provide its proper end for each thing. There was the centre of the universe as the end of motion for those things which are heavy, and the celestial spheres as the end of motion for those things whose natures lead them upwards. The celestial spheres were for things which are impassible and ingenerable, the lower regions for things impassible and generable. Nature was a drama in which each thing played its part." *Science and the Modern World* (New York: Macmillan Co., 1928), pp. 10–11 [ed. note updated]. It would be very desirable that the history of cosmic images be the object of a systematic study. The presence of a certain cosmic image in the spirit of an epoch does not settle any question in the philosophical order; nor does it affect in any manner the truth or error of a system of philosophy; but it can happen that it poses some psychological and pedagogical problems of primary importance. Those concerned in working for the diffusion of philosophical truth will act wisely if they inquire attentively into the cosmic images familiar to their listeners. It can happen that they put their seductive power in the service of error; in this case they ought to be fought on their own grounds, by means of other images. It can happen that they favor an adherence of minds to the truth; in this case one will be in a position to take precautions that the imagination not unduly facilitate the work of the intellect and not serve to conceal the insufficiencies of logic. For example, in the beginning of his work *Dieu: Son existence et sa providence* (Paris: Bloud et Gay, 1933), Gaston Rabeau remarks that the knowledge relating to the history of the earth has created in the public consciousness and in the works of philosophers a disposition unfavorable to idealism. Everyone considers as a well-established fact that our planet has existed through an incalculable number of centuries before there existed any organism capable of any knowledge. It is clear that such an image of the history of the world determines a strong repugnance to identifying being with its representation, and can contribute effectively to protecting minds

from the idealist error. But the idealists, who are subtle people, certainly have their own manner of giving an account of the notion of geological age, just as philosophical refutation of idealism continues to be as necessary as it was in the past, despite images which seem to render it superfluous. Let us observe that the study of cosmic images could supply precious teachings on the influence that theories of social ethics and theories relating to physical nature exert on each other. It has often been remarked that the ideas of modern evolutionists have their origin, not in the natural sciences, but in the social sciences (the optimistic philosophies of the eighteenth century, Condorcet, and the rest). Recovered by the natural sciences, in particular by the biological sciences, these ideas have soon reappeared in the social sciences with new traits and increased force. Moreover, it can be noted that these ideas have undergone parallel transformations in the two systems. During a certain phase in the history of these ideas, biologists and moralists represented the evolutionary processes under the aspect of quasi-continuous movements admitting only a minimum of jolts (Lamarckism; the reformist conception of social progress in classical democracy). Then a moment came when biologists took up the idea of abrupt change, of sudden mutation (Hugo de Vries, Thomas Hunt Morgan). In a parallel way, the visions of great catastrophes, substituting radical new forms for older forms without any transition, take on a much greater importance in the speculations on the development of societies (certain Marxist currents, revolutionary syndicalism). It would be easy to show that the vehicle of these reciprocal influences is nothing other than *the cosmic image*.

63. It would be useful to compare what Simon says here with the well-known views of Thomas S. Kuhn on paradigms and scientific revolutions. *The Structure of Scientific Revolutions* (Chicago: The University of Chicago Press, 1962). Both of them have clearly been influenced by the work of Émile Meyerson [ed. note].

2
Science and Systematic Knowledge

EACH OF US IS ACQUAINTED with Comte's sayings on predictions as the appropriate aim of scientific cognition. "Above all the scientist must predict," wrote Henri Poincaré,[1] but in a spirit quite different from Comte's. Citations could be multiplied attesting to the fact that according to the most widely held opinion in the last century and in our own day the prediction of phenomena is one of the essential functions of science, and the ability to allow foresight one of the distinguishing characteristics of scientific knowledge. There is an equivocation in that statement which the analyses of Georges Sorel, in an old work, largely unknown, but republished several years ago,[2] fittingly invite us to examine.

Sorel remarked that research about nature can be directed toward two quite different ends. Sometimes it is intended to predict phenomena with an approximation sufficient for the needs of action; sometimes it is intended to "connect scientific abstractions by laws independent of any possible conditions of real appearance."[3] The second approach is that of rational science; as to the first, it is characteristic of a discipline, quite far removed from the genuine scientific type, which Sorel designates by the felicitous term *systematic knowledge*. Let us state that in the thought of Greek rationalism, whose stringent demands Sorel seems clearly to have rediscovered here, the scientific object is constituted by the necessary relations between essences and their properties. Concrete reality being composed of nature and adventure, of lawfulness and chance, rational science knows only nature and lawfulness; its object is the possible rather than (concrete) reality, or, more precisely, the necessary formulation of abstract possibilities, the order and the logical sequence of *de jure* necessities.[4] The ordinary course of physical events in which countless interferences of causal series are produced presents a quite different aspect from the contingency-less world of pure science. Science, to the extent to which it conforms to its ideal, considers lawful processes in an *isolated* state,[5] and the statements that it utters concerning these processes *sheltered from chance* have only a strongly conditional predictive value. The pure theoretical formulation allows us to predict what will be the position or the state of the object considered at any given moment in the future on the condition that no disturbing influence intervenes. As to knowing whether in fact some disturbing influence will intervene in such a way as to modify the results, whether the eventuality of such a disturbance will be frequent or rare, the theoretical formulation has nothing to say at all. Concrete prediction is outside the realm of pure science.

But let us not conclude against the evidence that no concrete forecast is possible by the scientific way. A chance occurrence is not of itself unforeseeable.[6] Nor is it of itself resistant to precise calculation. And if the plurality of the initially defined data adequately corresponds to the plurality of factors at stake, the intersections of lawful processes could be calculated with as much rigor as the lawful processes themselves. This mode of foresight bearing on the interference of lawful processes, each of which is strictly determined, seems to be a characteristic of applied science. To take a simple example: the law of the expansion of mercury does not allow me to foresee unconditionally the elongation of the thermometric column which will correspond to a certain degree of heat; in fact, at the same time as the expansion of mercury causes the rise of the top of the column, the expansion of the tube, by increasing the volume afforded the liquid mass, tends to cause its descent. In order to foresee the point at which the top of the column will stop, it is then necessary to define two lawful processes and to calculate the result of their encounter. But the calculation of the result of their encounter, the scientific prediction of the concrete event, presupposes the exhaustive definition of the lawful processes at work and it is at this point that Sorel's criticism comes in.

"Physicists," he writes, "are quite far from being able to furnish constructors with the data which are necessary for them to take into account the phenomena occurring in the equilibrium of real bodies; they succeed only in treating very special cases of elasticity and preoccupy themselves with materials prepared only with their experiments in mind, materials having a much simpler structure than that of materials used in practice."[7] The simplification of laboratory materials aims at realizing the isolation of factors, at least in a relative sense, to bring the experimental material to a state as close as possible to the state of abstraction in which nothing would impede the factor under consideration from developing its properties and its proper causality, any disturbing encounter being excluded.[8] For lack of a perfect isolation which would confer a value of full and complete prediction on the pure theoretical formulation,[9] laboratory conditions reduce the scope of adventure and allow a relatively exact verification of the calculations,[10] thanks to a sufficiently complete inventory of the factors at hand. But in the complex conditions of everyday life, of nature not under the disciplined control of human initiatives, even of industrial practice, it is generally impossible, or exceedingly burdensome, to establish an exact count of the factors at stake. Then the mode of scientific prediction gives way to the mode of *systematic* prediction, not exactly as a consequence of the intervention of chance, but as a consequence of the eventuality of encounters between lawful processes, whose initial formulations cannot be exhaustively established. All in all, we have three modes of prediction: abstract scientific prediction bearing on an isolated factor, a highly conditional prediction that is the work of pure science; concrete scientific prediction bearing on an ensemble of factors and their interactions, a form of prediction of a less conditional kind, which would be absolutely exact in the extreme case in which the factors of the system would be exhaustively defined; and, finally, *systematic* prediction bearing on an ensemble, certain factors of which do not allow a scientific definition and can be included only in a practical solution. Of two doctors, one very learned, the other not much of a scientist but provided with long clinical experience, the less scientific will often make the more exact prognosis. It is not a question of disputing the usefulness of instruments provided by science in order to predict even the most complicated events. In a large number of cases, the conditional prediction exercised by scientific thought will be the main obstacle to an actually verified prediction. But if the superiority of non-scientific prediction over scientific prediction should be recognized in regard to events, if it happens that non-scientific thought that is well adapted to the study of concrete behavior is superior to scientific thought in the prediction of a concrete event, that should not bring any discredit upon science, for its proper function is not concrete prediction.[11] But Auguste Comte, who had never clearly distinguished between science and systematic knowledge, "rebelled violently against physicists bold enough to want to replace the simple and convenient Boyle's law with complicated and useless formulations."[12]

Sorel analyzes the procedures of systematic knowledge in passages that constitute a remarkable contribution to the theory of practical knowledge. While the scientific solution is related to the development of a lawful line—or of several lawful lines defined by the initial data—and sticks so *closely* to lawful functioning that any undefined interference entails its invalidation, systematic knowledge states the solutions that include at once the lawful processes, calculable encounters, and disturbances that cannot be predicted in a determinate way. Systematic knowledge allows us to definitely predict, or at least to reduce the chances of failure to a minimum, thanks to an *enclosure* sufficiently broad that the least predictable disturbances cannot compromise the result sought. The dimensions of this enclosure and of the field left for unpredictable disturbances are a matter of practical sense—an incommunicable quality resulting from personal experience—rather than of scientific understanding. "It is observable . . . that many circumstances cannot be scientifically defined; many causes are among those we usually place in the category of chance, such as accidental and hidden defects, the effects of bad weather, vibrations, and so forth. The practical solution should not be a mathematical solution, like the kind that a relation between geometrical magnitudes provides. It should *include* every solution that would correspond to all these indefinable conditions. *Practical sense* consists, above all, in the appreciation of the characteristics of enclosure; it is a sense that can be only acquired by a long and meticulous experience of things."[13] Following Sorel, the enclosure which is at issue can be constructed in two ways. Sometimes one proceeds empirically by comparing the findings obtained and the approximate definition of a proven model, to which it is prudent to remain closely attached. (This is *comparative* systematic knowledge.) Sometimes a theoretical system is set up, all of whose terms are connected by mathematical relations, and practical coefficients are added to this system. (This is *formal* systematic knowledge.) Sorel gives illustrations of these definitions by examples borrowed from architecture and industrial mechanics.

Does the generalized intervention of the notion of probability and the substitution of "statistical" formulations for "causal" ones make Sorel's analysis, which presupposes a rationalist conception of scientific lawfulness, out of date today? It does not seem that a probabilistic science has invalidated the regulative ideal of a rational grasp of necessities inscribed in the nature of things. Indeed, by modifying the ardor of reason, it teaches us, rather, to recognize at what level of scientific abstraction—doubtless more removed from the experimental datum than ancient physics believed—the requirements of this ideal are satisfied. Thus, it could be said that the question of scientific prediction, in the face of innovations introduced by contemporary physics, should be taken up at the point where Sorel left off.

Notes

1. *Value of Science*, p. 142. In Auguste Comte the theory of foresight as essential function of scientific knowledge is connected with a practical conception of the ends of science: *to know in order to predict in order to provide*. The ideal pursued by the theoretical path is not primarily theoretical, but demiurgic and ethical (human mastery over nature and the rational organization of societies). Henri Poincaré, on the contrary, has often declared in magnificent terms the primacy of the values of contemplation in the intention of the scientist. "If we wish more and more to free man from material cares, it is that he may be able to employ the liberty obtained in the study and contemplation of truth." Ibid., p. 11. Cf. also p. 137.

2. "L'ancienne et la nouvelle métaphysique," an article which appeared in *L'Ere Nouvelle* in 1894, was republished with an important preface by Édouard Berth under the title *D'Aristote à Marx* (Paris: Rivière, 1935).

3. Ibid., p. 97.

4. We find expressions about the scientific object very close to those we are using here in Edmond Goblot's *Essai sur la classification des sciences* (Paris: Alcan, 1898). Goblot concludes that science in the pure state would in no way be a picture of the world, but a picture of human reason. Aristotelian realism, at least in the form it takes in St. Thomas (thanks to the Augustinian theory of ideas), would rather see in it a

picture—infinitely inadequate—of divine reason; thus realist rationalism and idealist rationalism, in spite of their radical opposition, agree in recognizing only the *forms of reason* (divine or human) underlying the world of science. By considering things in this light it is easily explained why Marxist materialism, whose main characteristic seems to be the refusal to allow the presence of any idea at the basis of reality, qualifies the two great forms of rationalist thought as equally idealist.

5. The notion of an isolated system is precisely stated by Whitehead in the following terms: ". . . the conception of an isolated system is not the conception of substantial independence from the remainder of things, but of freedom from causal contingent dependence upon detailed items within the rest of the universe. Further, this freedom from causal dependence is required only in respect to certain abstract characteristics which attach to the isolated system, and not in respect to the system in its full concreteness." *Science and the Modern World,* p. 68.

6. Of course, it is a matter of prediction in an historical sense, not prediction in an essential sense. See Yves R. Simon, "Chance and Determinism in Philosophy and Science," in *The Great Dialogue of Nature and Space,* ed. Gerard J. Dalcourt (Albany, N.Y.: Magi, 1970), pp. 185–86 [ed. note].

7. *D'Aristote à Marx,* p. 97.

8. "The law is an ideal construction which expresses, not what happens, but what would happen if certain conditions were to be realized." Émile Meyerson, *Identity and Reality* (New York: Dover, 1962), p. 32.

9. At least in regard to the natural course of things for the possibility of a preternatural intervention by the First Cause cannot be ruled out in any case.

10. "The examination of hypotheses always leads us to discover in place of reality apparatus of human construction and functioning as those we use everyday. To pursue the question to the end, we should investigate whether there would not be some link between this way of constructing hypotheses and the experimental method, which uses apparatus constructed with so much precision that they could be regarded as being almost as perfect as geometrical figures." Sorel, *Les préoccupations métaphysiques des physiciens modernes,* p. 48. "Experimental science should be defined like this: To observe the mechanisms possessing geometrical characteristics to an eminent degree, removed from chance and in which a certain part of nature is included." Ibid., p. 52.

11. Sorel, *D'Aristote à Marx,* p. 113.

12. Ibid., p. 107. "Auguste Comte once decreed that one should guard against hypotheses and maintained that their usage belonged to the theological and metaphysical ages. Any progress would have been stopped if his contemporaries had taken him seriously, for never did one see a

man more stubbornly close his eyes to the new paths that were explored in his time. While Fresnel renewed physics by showing that the elasticity of ether should henceforth occupy an ever more dominant place in theories, Auguste Comte aspired to prohibit forming any hypotheses about the agents that produced phenomena and about their mode of action. It is notable that he rejected the idea of reducing optics to motion. What a calamity for science if physicists had read Auguste Comte's books! Fortunately they were only read by medical practitioners and that had no aftermath." Sorel, *Les préoccupations métaphysiques des physiciens modernes*, p. 36.

13. Sorel, *D'Aristote à Marx*, p. 99.

3
The Vienna Circle

SIGNIFICATION AND OBJECTIVITY

The following is an excerpt from a letter addressed to General Charles Ernest Vouillemin in response to one of his articles on the Vienna Circle and published after that article in the Revue de Philosophie (Paris) 1935.

READING YOUR ARTICLE gave me the feeling that at the very center of the conception of the Vienna Circle, there is a certain theory of *signification*. So many minds, among the philosophers and among the scientists, are satisfied with equivocation and imprecision that it is a joy to meet a group of researchers who in the forefront of their preoccupations maintain a basic concern never to utter a word without knowing exactly what they mean by it. The criticism of language is obviously the first and the most indispensable step in any scientific thought. No one will dare disagree with a principle so manifestly necessary. But when it is a matter of applying the principle, disagreement might be produced by reason of eventual divergencies concerning the theory of meaning.

If the current use of the term "sign" is considered, if the different categories of reality which present the characteristics of a sign under different headings are examined—an inductive procedure indispensable, hardly arbitrary, and in any case unassailable—one ends up with the following definition: a sign is that which makes something other than itself present to knowledge.[1] Nothing is the sign of itself. Every sign is essentially related to the object signified, which is the measure of the sign.

Sometimes the sign is a thing that is made known to the mind *before* making it know the object that the sign signifies. It is in this way that smoke is the sign of fire, that the flag is the sign of the country, that the letter A is the sign of the A sound. Before being effaced before the object signified, such signs show themselves in their own right. Let us call them *instrumental signs*.

Sometimes the sign is directly effaced before the signified and does not constitute an object known for itself, except through a secondary and reflective kind of cognition. Such is the situation, for example, when we have a memory, a sign of the object remembered. When I remember, my mind deals directly with a past event, thanks to a remembrance present to my memory. The remembrance becomes a known object only when I cease to consider the remembered event in order to reflect on the psychic impression that makes the event present to my memory. Signs of this kind are pure impulses of the mind toward the object, and because every sign is made to make an object present to the mind, because it is the function of the sign to efface itself before the object, one must say that remembrances fulfill the ideal of the perfect sign. We will call them *intentional* signs.

The concept, no less than the memory, is an intentional sign, as the distinction between science and psychology testifies. When the mathematician studies a certain mathematical object, the focus of his mind is on the object conceived or, better, on the conceived aspect of the thing known. When the psychologist studies the formation of mathematical concepts, he turns back from the object conceived to the mental sign that allows its conception. If we add that among instrumental signs some maintain a resemblance to the object signified—the portrait of a man is his image—while others maintain no relation of resemblance to the object they signify—the letter A is not the image of the A sound—we shall have reunited all the essential elements of a theory of meaning.

Words belong to the order of instrumental signs. Read or heard, they make themselves known as objects of visual or auditory perception before making known the object that they signify. But of what precisely are they signs? What do we intend to communicate when we say a word? By means of the word, we intend to generate the concept that we ourselves have in our minds in the minds of the persons with whom we are speaking. But we do not intend that our interlocutors stop at the concept, for, by causing the awakening of a concept by a word, we want to make them perceive the object that we ourselves perceive. It is this that the Aristotelian adage expresses: words are the signs of ideas which are themselves signs of things.

From all this it follows that the problem of meaning is the working out of the relation between the sign and the object signified. Insofar as it carries out its function, the sign is directed toward the object. In the fullness that only mental signs such as concepts, images, memories attain, the sign is objectivity itself. Insofar as it is a sign, the sign depends on its object and not on the knowing subject.[2] It depends on the knowing subject only insofar as it falls short in its function, when it signifies badly; when it is not a sign.

Consequently, the criticism of language, the verification of the meaning of words will have to follow a law of strict objectivity. Verifying the meaning of a word is, above all else, assuring oneself that the word signifies a concept and stating precisely the concept that it signifies. Secondly, it means making sure that the concept signifies a thing and determining what thing—or what aspect of what thing—it signifies. The second operation is the most exacting, and it seems that the authors that you cite, when they try to implement it, are moving in the direction of a kind of psychologism, a kind of subjectivism. If I understand your presentation correctly, according to them a term would be endowed with meaning when it corresponds to a mental act that I am capable of executing and testing, to a cognition lived in my subjectivity. Concerning meaning, the supreme instance would be a subjective *Erlebnis*. We believe, on the contrary, that the decision belongs to the object and to its evidence. The criticism of the oral sign, beginning with the definition of the concept represented by the word, will be developed by a resolution of the concept sought at the point where an object of absolutely primordial evidence is attained. According to the direction followed in the process of concept-resolution, the termination will be the evidence of the observable or the evidence of intelligible being. So long as one is not in a position to push this resolution to the end, one will not know exactly what one is saying. Every positive scientist is on the wrong track when he uses a term whose meaning cannot be reduced to observable data. Every philosopher is on the wrong track who uses a term whose meaning cannot be reduced to being, the primary datum of the intellect. The criticism of language terminates in the exclusion of metaphysics only if one decides to attribute the role of supreme instance to the *Erlebnis* rather than to the evidence of the object; for intelligible being possesses an evidence superior to any fragment of the sensible and the observable. "The empiricist does not say to the metaphysician: 'Your words assert something false', but 'Your words assert nothing at all!' He does not contradict the metaphysician, but says 'I do not understand you'."³ After this quotation from Schlick, you write: "In more familiar language, your discourse says nothing to me; I do not grasp your words. I do not understand, because in order to understand, I must relate everything to myself."⁴ I would say that I must relate myself and relate *my* signs to the object. It is the latter which will decide whether there is something to understand.

The New Empiricism

Rudolf Carnap, one of the best known members of the Vienna Circle, sums up his ideas on philosophy with remarkable clarity in the short work *Philosophy and Logical Syntax*.⁵

The account opens with a theory of verification. Logical analysis recognizes two kinds of propositions: those that deal with an object of present perception, and are verified by a present perception (e.g., "Now I see a red square on a blue ground"), and those that are not directly verifiable. A proposition P that is not directly verifiable "can only be verified by direct verification of propositions deduced from P together with other already verified propositions."⁶ Any proposition from which one cannot deduce any directly verifiable proposition is devoid of meaning; such is the case of metaphysical propositions. "Metaphysicians cannot avoid making their propositions non-verifiable, because if they made them verifiable, the decision about the truth or falsehood of their doctrines would depend upon experience and therefore belong to the region of empirical science."⁷ So the author admits from the outset without the slightest trace of justification that sensory intuition is the only source of certainty. Not for a moment does he ask himself whether certain propositions might be capable of being proven not by reduction to sensory evidence, but by reduction to rational evidence. Metaphysics is obviously excluded by the initial acceptance of the empiricist postulate, but no one tells us why we must accept this postulate.

Carnap is right to liken *problems of reality* to metaphysical problems: the theses of realism, of idealism, of positivism are metaphysical propositions devoid of meaning.

> Sometimes the views of the *Vienna Circle* have been mistaken for a denial of the Reality of the physical world [but we make no such denial]. It is true that we reject the thesis of the Reality of the physical world, but we do not reject it as false, but as having no sense, and its idealistic *antithesis* is subject to exactly the same rejection. We neither assert nor deny these theses, we reject the whole question.[8]

What does ethics become in such a conception of knowledge? Often we designate under the name moral sciences certain disciplines pursuing empirical investigations about human actions; these are empirical disciplines that in no way belong to philosophy. As to normative ethics it does not formulate (true or false) propositions, but expresses wishes that are neither true nor false, but efficacious or inefficacious, and contain no theoretical meaning. "But actually a value statement is nothing else than a command in a misleading grammatical form. It may have effects upon the actions of men, and these effects may either be in accordance with our wishes or not; but it is neither true nor false. It does not assert anything and can neither be proved nor disproved."[9] It is useless to insist on the scope of these theses. It has often been noted that the Vienna Circle is devoid of the spirit of scientific imperialism which not long ago animated the supporters of scientism; its members are supposed as least likely in the world to want human life to be subject to the dictatorship of positive science. But look at what price this liberation is purchased. Ethics is emptied of its theoretical soul, and no longer do the laws of human conduct take their form at the source of truth.

The proscription of metaphysical propositions as devoid of meaning raises an objection of a psychological order. Metaphysicians have existed in every epoch, and reading their works produces an effect, sometimes a strong one, on the reader. How would that be possible if metaphysical propositions do not express anything? Carnap answers, "they *do* express something, but nevertheless they have no sense, no theoretical content."[10] Metaphysics is assigned to the *expressive* function of language, not to its *representative* function. It has a kinship with lyric poetry; metaphysical systems express emotional makeups. Carnap writes elsewhere that "Metaphysicians are musicians without musical ability."[11]

But while lyric poetry and music confine themselves to their expressive function, metaphysics gives the appearance of having a theoretical content; in doing so, it is misleading, and this is why it should be rejected. Let us be grateful to Carnap for having said so well what he thinks or what so many amateurs who cultivate metaphysical systems for their human expressive value, without being concerned with their theoretical value, would think, if they were sincere. They are musicians whose absence of musical talent is a matter of regret.

Metaphysics being relegated to the expressive function of language, psychology to the order of the empirical sciences, ethics on the other hand being divested of any theoretical meaning, *logical analysis* absorbs all that can have validity in philosophical thought. We cannot reproduce the extremely compact description that Carnap gives of logical analysis. Let us only point out that his thought seems resolutely aimed in the direction of an absolute logical formalism.

Hans Hahn's pamphlet "Logic, Mathematics and Knowledge of Nature"[12] makes a first-rate contribution to the exposition of the new empiricism from a perspective quite different from Carnap's. We know that Hans Hahn, who died prematurely, is considered to be the true founder of the Vienna Circle. He starts off with the simple remark that there are two kinds of statements in physics: those that concern simple observation (if one plucks a taut string, one hears a sound), and those that presuppose theoretical bases (a hydrogen atom consists of a positively charged nucleus around which a negatively charged electron revolves). Thus, the eternal problem of the relations between observation and theory is found laid down. Hahn sketches the dialogue between philosophers in the following way: rationalism, born of the criticism of sensory illusions, holds that thought alone is certain; early empiricism (Locke and his successors) avails itself of the results of modern physics, but shows itself incapable of explaining the conformity of logical or mathematical propositions with reality. For a dualistic conception, "*Thought* grasps the most general laws of all being . . . ; *observation* provides the detailed filling of this framework."[13] Finally, following the *usual* conception,

> we learn certain facts, which we formulate as "laws of nature"; but since we grasp by means of thought the most general lawful connections (of a logical and mathematical character) that pervade reality, we can master a knowledge of nature by basing it on facts borrowed from general observations to a much larger extent than we could have just by observation in the proper sense of nature itself. . . . the experimental physicist provides knowledge of laws of nature by direct observation.
>
> The theoretical physicist thereafter enlarges this knowledge by thinking. . . . This view seems to be strongly supported by numerous discoveries that have been made with the help of theory, like—to mention just some of the best known—the calculation of the position of the planet Neptune by Leverrier, the calculation of electric waves by Maxwell. . . .[14]

This usual conception is entirely untenable. Hahn sees a mystical and theological representation in it, presupposing a pre-established harmony "between the course of our thinking and the course of nature."[15] His conclusion is that we should return to a purely empiricist standpoint. "However, we shall have to avoid the error committed by earlier empiricists, that of interpreting the propositions of logic and mathematics as mere facts of experience."[16]

In the author's view logic is not concerned with any object, but only with our way of speaking about objects. A plant having a certain description bears the name "snow rose"; it is also called "helleborus niger." The statement "every snow rose is a helleborus niger" is endowed with absolute certainty and unlimited universality, but it expresses nothing more than a *convention* concerning the way in which we designate this plant. "Similar considerations apply to the principles of logic."[17] So, when we say "no object is both red and not red," "every object is red or not red,"

> These two propositions, the law of contradiction and the law of the excluded middle, say nothing at all about objects of any kind. They do not tell me of any of them whether they are red or not red, which color they have, or anything else. They merely stipulate a method for applying the designation "red" and "not red" to objects, i.e. they prescribe a *method of speaking about things*.[18]

But we are not told why we would want to speak in this way. The obviously conventional choice of verbal signs aside, would it not be because we are constrained to do so by the object?

Thus, logical deduction never results in anything other than tautologies: "our conventions regarding the use of the words 'not' and 'or' is such that in asserting the two propositions 'object A is either red or blue' and 'object A is not red', I have implicitly already asserted 'object A is blue.'"[19] Logical deduction "makes us aware of all that we have implicitly asserted—on the basis of conventions regarding the use of language."[20] If we have recourse to logical deduction, it is that, not being omniscient, we are not immediately aware of all the statements in a relationship of tautological equivalence with the one that we have uttered. Mathematical statements, like logical statements, are tautological. Between 2 + 3 and 5 there is exactly the same relation as exists between *Helleborus niger* and snow rose.

This whole argument is far from being clear. On the one hand, Hahn asserts that deduction is reduced to a tautological transformation and only makes us aware of the equivalence of certain linguistic conventions; on the other, he asserts that deduction is justified and made necessary by the fact that we are not omniscient. What, then, is this science that we do not yet entirely possess, and whose imperfection forces us to have recourse to deduction? If it is true that the discovery of equivalencies between linguistic conventions provides a remedy for our lack of omniscience, we would have to admit that this omniscience concerns words and not objects. For instance, when I learn that the word *Pferd* has the same meaning as the word *horse*, I make no progress in the cognition of the object designated by these two words in different languages, and if the German-English dictionary makes up for my lack of omniscience, that is only a matter of a lack of linguistic omniscience. I make progress of a quite different kind and of a quite different value when I learn that $24 \times 31 = 744$ by an operation of multiplication.

This elementary observation prompts us to give a precise formulation of the notion of tautology. Unless we are mistaken, a tautological statement is one in which there is no kind of distinction, other than eventually a verbal one, between the subject and the predicate. Such a statement informs us only about the equivalency of linguistic conventions (Bratislava is Pressburg). On the contrary, every time there is any distinction whatsoever between the object signified by the subject and the object signified by the predicate (even if there is an identity between the subject word and the predicate word, as is the case when we say *being is being*), the statement is not tautological, and it constitutes, if it is well-founded, progress in the knowledge of the object. Now, for there to be more than a verbal distinction between the subject and the predicate, it is not necessary that there be a real distinction, so far as the object is concerned, between what is envisaged by the subject and what is envisaged by the predicate; it is enough that the subject and the predicate correspond to aspects of reality conceptually grasped as distinct, that they correspond to objectively distinct concepts, to distinct formalities, even if these aspects of reality, these objective concepts, are in fact identical and actually imply each other.[21] Between the object of thought 24 × 31 and the object of thought 744, there is obviously no real distinction. Moreover, the object 24 × 31 actually implies its identity with 744. But this identity will be revealed to me only by a calculation, whose function will have consisted in making explicit an identity of objective concepts which was only implicit. The lack of omniscience that prevented me from seeing the identity of 24 × 31 and 744 at first glance deals not only with the equivalence of linguistic conventions but also with the identity of objective concepts. And that is why mathematical deduction advances our knowledge of objects, and not only that of the verbal signs which represent them. And why it is fruitful and not tautological, even though no real distinction intervenes between the objective terms that the deduction joins in its conclusion.

We see how the new empiricism intends to remove the difficulty in which the earlier empiricism had failed. Logical and mathematical statements are neither experimental statements nor synthetic propositions *a priori;* they are tautological statements. But if it is true that thought gives nothing more than what observation gives us, how is one to explain the success of the theoretical track, such as the discovery of Neptune? Hahn's description of the process of this discovery deserves to be closely followed. In the first place we are aware that what we know of the phenomena of motion "can be well described in a unified way by the assumption that between any two mass points a force of attraction is exerted which is proportional to their masses and inversely proportional to the square of their distance."[22] It is only an assumption. It is impossible to affirm a law since it is impossible to observe the behavior of all the mass points. Now, in formulating this hypothesis, one formulates without knowing it a number of other statements whose calculation will make us particularly aware of the statement (brought out by Leverrier's calculations) that a planet at a definite place in the heavens must be visible. Observation confirms the hypothesis, and it is observation alone which establishes the existence of the planet. Observation could have shown that nothing was visible at that place in the heavens and invalidated the hypothesis. That, moreover, is what actually happened later: "in asserting the law of gravitation, one implicitly asserts that at a certain time the planet Mercury must be visible at a certain place in the heavens."[23] But observations showed that it was not visible at exactly the position indicated in the heavens, and Newton's theory was replaced by Einstein's.

If every science must be verified by experience, it seems that a scientific statement should use only words "capable of being based on the observable." Hahn recalls that Mach was critical of the use of words like molecules and atoms "which are not directly attributable to what can be observed." According to Hahn, Mach's requirements would lead to a complete overthrow of science. It is sufficient to remark that the word "every" cannot be constituted on the observable, because it entails no corresponding observable object. It is necessary to learn to discern among the statements containing words that are not "constitutable" those that are legitimate and those that are not legitimate.

A word not "able to be constituted" has a legitimate use when it is accompanied by rules indicating how the statement in which the word occurs can be transformed into a statement containing nothing more than words "able to be constituted." "For instance, the kinetic theory of gases starts out with propositions about the behavior of molecules in order to end up, thanks to appropriate transformations and interpretative rules, with statements about the behavior of real gases and verifiable by observation, in which the word molecule no longer occurs."[24] Statements containing only words "able to be constituted" are comparable to gold certificates; statements containing words which are "unable to be constituted," but are statements convertible into those containing only words "able to be constituted" are comparable to checks backed with sufficient funds. But if this convertibility does not exist, the statement is comparable to a check not backed with sufficient funds, and such is the case with metaphysical statements.

This entire analysis seems very remarkable to us, and we are pleased to note that it retains its value outside the empirical perspective in which the author pursues it. Let us recall the passage in *The Degrees of Knowledge* in which Jacques Maritain shows what distinguishes the *invisible* for the scientist from the *invisible* for the philosopher:

> But the moment the scientist passes to an order [the atomic order] in which the very possibility of full and continuous observation of phenomena is eliminated, he passes from a world of objects imaginatively representable to a world of objects without any imaginable form. Such a world is unimaginable by default, or "privatively."[25]

This "unimaginable by default" remains "indirectly observable." By contrast the philosopher goes from the visible "to what is *of itself* outside the order of sensible observation (for the simple reason that the principles which he reaches are in themselves pure objects of understanding and not objects of sensible apprehension or imaginative representation. Here is a world unimaginable by nature, or 'negatively')."[26] The "directly observable" of Maritain corresponds to Hahn's words "able to be constituted." Maritain's "indirectly observable," "privatively unimaginable" correspond to the words "unable to be constituted," but in such a way that the statement in which they occur can be reduced to a statement containing only words able to be constituted. The words "unable to be constituted" are such that the statement in which they occur cannot be reduced to a statement containing only words "able to be constituted"—and metaphysical statements cannot be so reduced—words that correspond to Maritain's "unimaginable by nature or negatively." The whole question is to know whether the "unimaginable by nature" is the object of a valid cognition, as critical realism thinks, or can only give rise to an illusion of cognition, as Hahn and the members of his group assert.

The last chapter is devoted to the problem of truth; it is particularly revealing about the philosophical tendencies of the Vienna Circle. Hahn points out that the old conception of truth as conformity of a proposition with what happens in the real world cannot be retained because the real world is beyond our range. The *pragmatic* conception of the truth remains and Hahn adheres to it: "the truth of a proposition [consists] in its confirmation."[27] A statement is true insofar as the prognostications that it implies are confirmed. That is a position which does not seem to go beyond the kind of *methodological pragmatism* currently practiced by scientists and which, we believe, cannot affect in any way the metaphysical notion of truth. But Hahn considers Poincaré to be mistaken when he declares that physical science is essentially distinct from historical science in that it makes prognostications. The proposition "John Lackland passed by here" is also in the final analysis a prognostication or, more exactly, a piece of information, an indication for making prognostications, one that can be confirmed or not, just like a physical proposition. It concerns prognostications of roughly the following type: when competent men start a more intensive study of the sources available to them, and then discover new sources, they will not fail to repeat that "John Lackland passed by here." Thus historical truth and scientific truth are of the same kind; *unitary science exists.*

But then why will competent men repeat with a growing assurance that John Lackland did pass by there? Is it because they foresee that other men coming after them, even more competent than they are, will repeat the statement with even greater assurance? Or is it because they suspect that John Lackland really passed by there?

The pragmatic conception of truth applied to history as the confirmation of a prognostication gives rise to jests quite apt to make us understand why this conception, not only of history, but also of science, must be rejected. For if it is true that to confer meaning on the proposition "John Lackland passed by here" and indulge in any set of prognostications in its regard, it is necessary to recognize that this proposition is measured by an historical *being,* an actualized and unrepeatable existence, then is it not just as necessary in order to confer meaning on the statement of any law whatsoever, and to construct prognostications based on this statement, to admit that it is measured proximately or remotely by the reality of a universal principle capable of supporting repeatable existential situations?

Philosophical-Scientific Illusions

Concerning a book by Philipp Frank, La fin de la physique mécaniste *(Paris: Hermann, 1936).*

This is a polemical work aimed at destroying illusions created in the minds of certain philosophers by the new physics. Though he is incapable of correctly formulating the philosophical problems to which he alludes, Frank characterizes the new forms of physical science in a most felicitous way. In this regard he deserves the respect of philosophers.

Some people who were too busy, too naïve, or too clever wanted to believe that *the end of mechanistic physics* meant the discrediting of mechanistic philosophy and opened up new perspectives for vitalism, spiritualism, and the affirmation of freedom. Frank tries to put the revolution that has occurred in the general conceptions of physics since the beginning of the century in proper perspective. While classical physics asserted the universal validity of the Newtonian formulas, the new physics recognizes that other formulas must be substituted for them when the bodies under investigation are moving at very great velocities (Einstein) or involve very small dimensions (Planck). Will one say, then, that the new science is more "spiritualistic" than the old for the reason that it deploys a more complex and refined mathematical apparatus? I recommend to the reader the following remarks of Frank which seem to me full of meaning and good sense: "If we seek the difference between the old and the new physics, we find that they both are made up of mathematical formulas, and that there are methods in each of them allowing for the establishment of relations between mathematical dimensions that occur in the formulas and the observable dimensions that are the result of measurements."[28] "If mathematics contains a 'spiritual' element, it was also contained in classical mechanics."[29] "The difference between them rests solely in this fact . . . that the formulas of classical mechanics are no longer relevant for very fast bodies or for very small ones, while the formulas of the new physics remain useful in both cases."[30] Frank's views can be summed up by stating that in the mathematical reading of physical reality, contemporary scientific disciplines have substituted an instrument that works for all velocities and all dimensions for an instrument that worked only for reduced velocities and for macroscopic dimensions. Such a substitution takes place completely within the field defined by the general principles of mathematico-physical science, and entails no basic innovation in the philosophy of nature.

The greatest portion of this little work is concerned with the question of determinism and free will. We would suggest that the reader disregard the philosophical poverty that weighs down Frank's presentation and pay attention to several pages in which he expresses with a particular precision the mathematico-physical meaning of the notion of determinism. The ontological conception of nature being alien to him, Frank shows perfectly from the point of view of the physicist that the notion of determinism makes sense only in relation to a possibility of exact calculation.[31] A motion is said to be determined when it is possible, given a system of initial data, to predict the spatio-temporal coordinates of the moving object at a definite moment. The idea of indeterminacy expresses nothing more than the (essential) impossibility of such a prediction. When we understand that mathematico-physical science views things under the aspect of the measurable, as philosophy views them under the aspect of being, when we understand that the terms that philosophy and mathematico-physical science have in common are recast by the latter in such a way as to signify something in relation to the possibilities of measurement analogous to that which they signify elsewhere in relation to being, we are tempted to judge as childish the emotion aroused in philosophers by the emergence of indeterminacy physics. In philosophy we describe something as necessary which cannot be other than it is, and we describe something as contingent which can be other than it is. In mathematico-physical science, we describe the event capable of being predicted by an exact calculation as determined, and the event which—for essential reasons— eludes such a prediction is described as undetermined. There is an analogy between these pairs of notions: *the second is related to the measurable as the first is related to being*. But these notions do not overlap. A chance event in the philosopher's meaning of the term may in the physicist's sense be predictable and determined. When considering the discussions relating to these problems, one has the feeling that even with the philosophers properly philosophical notions have been abolished in favor of scientific notions sophistically burdened with a confused human resonance. If this were not so, how could the problem of freedom have been brought up in connection with this account of electrons and energy barriers?

Notes

1. See Jacques Maritain on the theory of the sign, "Sign and Symbol" in *Ransoming the Time* (New York: Gordian, 1972), pp. 218–26. Abundant quotations from the admirable dissertation of John of St. Thomas on this subject (*Logic II*, questions 21–22) will be found in the notes.

2. Except, obviously, in what concerns existential conditions.

3. See Moritz Schlick, "Positivism and Realism," in *Essential Readings in Logical Positivism*, ed. Oswald Hanfling (Oxford: Blackwell, 1981), p. 110 [ed. note].

4. Général Charles Ernest Vouillemin, "Le positivisme critique de l'École de Vienne," *Revue de Philosophie*, n.s. 6, No. 3 (1935), 239.

5. *Philosophy and Logical Syntax* (London: Kegan Paul, 1935).

6. Ibid., p. 11.

7. Ibid., p. 17.

8. Ibid., pp. 20–21.

9. Ibid., p. 24.

10. Ibid., p. 27.

11. "The Elimination of Metaphysics" in *Logical Positivism*, ed. A. J. Ayer (Glencoe, Ill.: Free Press, 1959), p. 80 [ed. note].

12. In ibid., pp. 147–61. The text was translated by Arthur Pap, but it includes only parts of the original German work [updated note—ed.].

13. Ibid., p. 150.

14. Ibid., pp. 150–51. [The translation has been altered—ed.]

15. Ibid., p. 151.

16. Ibid., p. 152.

17. Ibid., p. 153.

18. Ibid.

19. Ibid., p. 156.

20. Ibid., pp. 156–57.

21. Three kinds of distinction, that of reasoning reason (a distinction between signs only), that of reasoned reason, and virtual distinction which corresponds to the latter as far as the object is concerned, are discussed in *Material Logic of John of St. Thomas*, pp. 76–88.

22. Ibid., p. 160.

23. Ibid., p. 161.

24. Hans Hahn, *Logique, mathématique et connaissance de la réalité* (Paris: Hermann, 1934), p. 44. [The original title was *Logik, Mathematik, und Naturerkennen* (Vienna: Gerold, 1933), ed. note].

25. P. 47.
26. Ibid.
27. Hans Hahn, *Logique, mathématique et connaissance de la réalité* (Paris: Hermann, 1934), p. 47.
28. Philipp Frank, *La fin de la physique mécaniste* (Paris: Hermann, 1936), p. 15. [There appears to be no English version of this work, ed. note.]
29. Ibid., p. 16.
30. Ibid., p. 15.
31. Frank affiliates himself with the school of physicists who pursue "an entirely disontologized science." See Simon, *Great Dialogue of Nature and Space,* p. 201 [ed. note].

4
Epistemological Pluralism

IT IS BANAL TO POINT OUT that modern epistemology, to the extent especially that it remains faithful to the Cartesian ideal, opposes a monistic conception of knowledge to the pluralistic ideal that was characteristic of Aristotelianism. What is perhaps less noticed is that a more or less acknowledged epistemological monism often subsists within a pluralistic gnoseology. (I am using the term epistemology to mean the theory of science; by gnoseology I mean the more abstract and more general theory of knowledge). If we take scientism as it is usually typified, the extreme scientism whose strict representatives have without doubt been rare—at least among minds of any breadth—we must say that here epistemological monism becomes identical with an absolute gnoseological monism: all scientific knowledge proceeds or tends to proceed from a univocal intellectual light which is that of positive reason. Here we see the first phase of the scientistic approach. It matters little, from the point of view we are taking, whether concretely we conceive of positive reason in a more empirical or a more mathematical manner; what is essential is the univocity of the notion rather than its content. But scientism also claims to reabsorb into a univocal science all certain knowledge: apart from such a science there could only be conjectures, precariousness, and fantasy. This means we have to give up religion and metaphysics and let positive science take over the domains of psychology, sociology, ethics, economics, and political science.

If now we consider the reactions that have arisen in the last two generations against scientism, we note that most of them, with a considerable diversity of terminology, have in the end limited themselves to affirming the validity of certain kinds of knowledge that cannot be subsumed under positive science. The validity of religious knowledge, of metaphysical knowledge, of moral knowledge has been ardently proclaimed. But to none of this extra-or supra-positive knowledge is attributed a scientific character; this remains reserved for what we call positive science, and thus scientistic monism is denied the second part of its claim but granted the first.

The development of Bergson's ideas seem to us to bear a rather remarkable testimony to the persistence of bias in favor of the univocal in modern epistemology. One could say that the entire gnoseological aspect of Bergsonism can be epitomized as an admirable effort to escape univocity, but Bergson's intuition escaped the framework of scientistic univocity only by resolutely placing itself outside the perspectives of demonstrative knowledge. Even though Bergsonian gnoseology may be pluralistic or at least dualistic, Bergsonian epistemology remains monistic. Consider also the attitude of the Vienna Circle: it manifests itself as far removed from the imperialistic spirit that characterized the heyday of scientism. "The groups represented here," declared Philipp Frank at their congress in Prague in 1934, "are the last to overestimate the importance of science for life."[1] This is fine, but the speaker goes on to say: "We know perfectly well that human development is determined more by instinctive tendencies than by openly scientific ideas."[2] Thus, unitary science—this is the very term which these Viennese use—is concerned with only a limited field in our universe, but what is outside this field is given over to the domination of the instincts.

In our view, whoever wants to work out a theory of the relations between philosophy and the sciences should above all take note of the scientific character of philosophy and understand that metaphysics, which is the archetype of all philosophical thinking, is at the same time purely and simply the archetype of all scientific thinking. If we often have to oppose science—in the restricted and modern sense of the word—and philosophy, we ought to do it with full awareness of what it involves and without ever losing sight of the fact that the most truly scientific sciences are of a philosophic sort. There are without doubt admirable analogies between philosophy, art, and religion, but far from testifying against the scientific character of philosophy, these analogies testify in its favor. In fact, if the philosopher experiences some affinity for the artist or the religious man, it is because like them he spends his life in intercourse with mystery.³ Socrates's dictum remains always true: the great superiority of the learned over the ignorant man is that the learned one knows that he does not know, and the more science realizes its ideal by going more deeply into its object, the more it becomes aware of its inadequacy and the more it acquires a feeling for mystery, through a marveling which resembles that of the artist and foreshadows in a dim way the obscure face-to-face vision of mystical experience.

We have observed that in scientistic epistemology, be it of an absolute or a limited sort, the exclusion of philosophy properly so-called or at least the negation of its scientific value is tied in with a univocal conception of scientific knowledge. If knowledge, wherever it is distinct from existence, that is, everywhere except in God, is a reiteration of being, then perfect knowledge, science, will be a perfect reiteration of being, and therefore it ought to admit of as many different types as being itself does. Grant the realist conception of knowledge as the reiteration of being and grant further the fundamental doctrine of the analogy of being, you are at the same time granting the principle of epistemological pluralism. People have not paid enough attention to the import of the pluralist principle in the Thomistic conception of science, and too many believe that all has been said when one has pointed out that for Aristotle and St. Thomas the speculative sciences involve degrees of abstraction that are irreducible. If I may be permitted to introduce here a theological allusion, I would remind you that according to St. Thomas the most perfect human knowledge that we can conceive of in the state of our present life, the knowledge of Christ, is infinitely diversified: the human intellect of the Christ has as many scientific lights as there are essences to know. At just about the time when Descartes was writing his celebrated page of the *Rules* on the unity of our natural light, John of St. Thomas was again pointing out, at the beginning of his treatise on the division of the sciences, this great ideal of a multitude of sciences coinciding exactly with the multitude of scientific objects.⁴

We mentioned a moment ago that modern epistemology, or, if one prefers, modern science in its theoretical exposition, *in actu signato,* has in general remained faithful to the monistic ideal of Cartesianism. Should we, on the other hand, consider modern science as lived, *in actu exercito,* it would seem that the evolution of science in the last three centuries presents on the whole the picture of a process of differentiation which continues without letup. It is as though the movement of history were dragging scientific thinking more or less successfully toward the pluralistic ideal realized in the human intellect of the Christ. Thomists ought to be the first to rejoice over this event.

If we compare today's differentiated knowledge with the system of the sciences such as St. Thomas conceived of it, what above all strikes one's attention is the dissociation of the science of nature from the philosophy of nature. It is above all on the first level of abstraction, on the level of physical abstraction, that the process of differentiation makes itself felt. The distinction between the science of nature and the philosophy of nature seems to us to be a definitive acquisition of scientific and epistemological thought. Let us hasten to note, however, that this distinction does not itself in any way preclude ulterior differentiations. To affirm the distinction between the science of nature and the philosophy of nature is in no way to prejudge the question of the specific unity of the two sorts of sciences thus designated.[5]

It is most instructive to try to uncover the reasons for all the resistance that this fundamental division of physical knowledge arouses from numerous Thomists. Until recently, modern Thomists had the habit of incorporating the philosophy of nature into metaphysics, under the names of cosmology and psychology. It is not necessary to point out that from the point of view of St. Thomas's doctrine this assimilation of physical and metaphysical philosophy cannot be maintained for a moment. But is it merely a question of being unfaithful to St. Thomas? To our mind, it is a question of a confusion of viewpoints that is extremely prejudicial to the philosophy of nature, destructive to metaphysics, and well suited to make definitively obscure the problem of the relations between philosophy and the sciences.

Other Thomists, no matter what they think of the distinction between the philosophy of nature and metaphysics, find it repugnant to admit the validity of a nonphilosophical science of nature. One of them told me once that in his view modern science is a false philosophy, so that it would be better, if we had the time, to substitute for it a true philosophy; in concepts of the empiriological type, to go back to Maritain's terminology, this author saw only pseudo-concepts, bundles of images. This extremely interesting attitude, which we could denominate an ontological integralism, presents some striking analogies with political clericalism. It is clear that it is, above all, in ontological notions that the intelligibility of being and the magnificence of truth manifest themselves, but does it follow that intelligibility is lacking in every nonontological apprehension of things? Likewise, it is evident that it is in a spiritual community that Christian life obtains the fullness of its bloom; does it follow that no temporal state can be Christian? Ontological integralism will vacillate between two attitudes. Sometimes it will attempt to violently take over areas that do not belong to it, and then we have philosophical pseudo-explanations of issues quite unsuited to philosophical explanation; we find a few examples of such pseudo-explanations in Aristotle and more numerous examples of it in the decadent scholastics, in Hegel, and in the romantic *Naturphilosophie*; such conduct is analogous to clericalism's, as it seeks to absorb the temporal into the spiritual. Sometimes ontological integralism encloses itself in a splendid isolation and abandons to the imagination and to interior and utilitarian mental functions the universe whose conquest it has renounced, a bit like certain contemporary theoreticians who abandon to the devil terrestrial affairs and condemn as blasphemous the notion of a secular Christianity. In opposition to ontological integralism we have to make clear that despite the obscurity of empiriological notions they can provide a high degree of certainty. When I say "dog" or "cat," I am thereby expressing a true concept, even though it is obscure and cannot be made precise except by evoking sensations. In those passages of his writings in which he shows that a concept is fundamentally different from an image because of its essential relationship to being, Garrigou-Lagrange declares that he will borrow his examples, not from the level of the inductive sciences, but from the level of the sciences that attain their objects in their essences, no matter how progressively and incompletely they may do so, viz., mathematics and philosophy. Pedagogically this procedure is well justified; one should always start with the clearest cases. It would be proper today to carry out the complementary task of showing that the ideas that share the least in any metaphysical character and that are the most tied in with the data of the senses and the imagination remain, they too, essentially relative to being. Although the bond that connects thought to being is more tenuous here than elsewhere, it is not broken and so long as any authentic science exists, it remains possible.

The division of physical knowledge into science and the philosophy of nature will, obviously, be rejected by all those who proclaim the vanity of ontological thinking. Because it seems to me to be the exact counterpart of the error we have qualified as ontological integralism, I shall cite especially the view put forth by Rudolf Carnap in his pamphlet *The Elimination of Metaphysics Through Logical Analysis of Language*.[6] The terms that we qualify as empiriological, for instance, "arthropod," Carnap declares to be meaningful, because they are reducible to observations; the terms we describe as ontological Carnap declares devoid of meaning because no observation could ever uncover the object they signify. What is lacking here is the intuition of being, and this is extremely serious. But if we brought to light the persistence of a bond between the intelligence and being in empiriological thought, we would at the same time be showing that the validity of empiriological thinking (which turns away from being even while allowing itself to be sustained by it) absolutely requires the validity of metaphysical thought (which devotes itself with all its strength to being), so that if the notion of being has no meaning, the notion of arthropod will not have any meaning either and as a result all science, along with metaphysics will theoretically be reduced into a substitute for music.

The term "epistemological pluralism," which we have used, is justified by the analogical character of being, the object of all scientific knowledge. Wherever we find a unity that is only analogical, predominance reverts to the pluralistic aspect. But because analogical relations preserve equally well a relative unity in the diversity of being, epistemological pluralism could never be an absolute pluralism. Through it are excluded all systems of strict compartmentalization or of multiple truth. Just as we cannot establish a bulkhead between philosophy and faith, so we cannot place one between science and philosophy. And should it happen that materially identical propositions be true from the point of view of the scientist and false from the point of view of the philosopher or vice versa, it will still be necessary that the diversity of points of view be justified in the unity of a superior view which can only be philosophical.

Notes

1. "Berichte: Die Prager Vorkonferenz 1934," *Erkenntnis,* 5 (1935), 5.
2. Ibid.
3. On the meaning of mystery in philosophy compare Jacques Maritain, *A Preface to Metaphysics* (New York: Sheed & Ward, 1939; repr. New York: Books for Libraries, 1979), First Lecture; Gabriel Marcel, "On the Ontological Mystery," in *The Philosophy of Existentialism* (New York: Citadel, 1963), pp. 9–46; Réginald Garrigou-Lagrange, *Le sens du mystère et le clair-obscur intellectuel: Nature et surnaturel* (Paris: Desclée de Brouwer, 1934). On the same subject let us mention this piquant passage of Georges Sorel's: "People have often laughed at Hegel's belief—that humanity, since its origins, had worked to give birth to the Hegelian philosophy, and that with that philosophy Spirit had at last completed its development. Similar illusions are found to a certain extent in all founders of schools; disciples expect their master to close the era of doubt by giving final solutions to all problems. I have no aptitude for a task of that kind. Every time that I have approached a question, I have found that my enquiries ended by giving rise to new problems and the farther I pushed my investigations the more disquieting these new problems became. But philosophy is after all perhaps only the recognition of the abysses which lie on each side of the footpath that the vulgar follow with the serenity of somnambulists." *Reflections on Violence,* trans. T. E. Hulme (New York: Collier, 1961), pp. 29–30 [note updated, ed.].
4. "The knowable object is a complex whose components are a subject and the property or affection attributed to it by demonstration. The object constituted by the truth of the inferred conclusion must be illuminated by a middle term contained in the premises. The middle term is the principle from which the concluded truth is inferred and by which it is illuminated in the very act of being inferred. Thus, the definitions which play the part of principles or middle terms in the demonstration of properties must be of such character as to determine the knowability of the inferred object by illuminating it. If quiddities were known adequately and according to what they are in themselves, each of them would determine, with regard to its properties, a science distinct from all other science. St. Thomas holds it probable that in Christ distinct infused sciences correspond to the diversity of the ideas representing quiddities (iii. 11. 6). However, in our present condition, the intellect proceeds by uniting and disengaging, fails to understand adequately any nature as it is in itself, co-ordinates natures with each other, links one nature to another, and, conversely, understands one and the same thing in diverse ways. The result is that diverse natures fall under the same science, whereas one and the same nature is studied by diverse sciences.

We have to explain how one principle can unite several natures in one science and how the study of natures can be divided into several sciences by a multiplicity of principles." *Material Logic of John of St. Thomas,* pp. 552–53. "If things are considered materially and in their real existence, a science enjoying the unity of an ultimate species treats of things that belong to various species. This happens obviously in metaphysics, in logic, in physics, and in any other science no matter how specific its unity. Just as there is no distinct power of vision for each species of color, so there is not a distinct science for each species of things. The specific distinction of the sciences cannot proceed from the distinction of the quiddities or things in their existence as things. The reason for this state of affairs is that our sciences, which are imperfect, do not coincide absolutely with the things themselves and do not comprehend them adequately. Each thing, if perfectly comprehended, could constitute the foundation of a science proper to it and specifically distinct from any other. Science would not require a co-ordination of ideas, and each thing, perfectly represented by its own adequate idea, would need no more than its idea to demonstrate its properties. From this consideration, St. Thomas deduces, in a firm argumentation (iii. 11. 6), that there was in Christ our Lord an infused science diversified into a multiplicity of habitus, for Christ understood the natures of things by ideas as diverse as the things themselves. His infused science did not consist in a multitude of ideas co-ordinated by one way of knowing and one kind of immateriality. Each idea, because it represented each thing perfectly and adequately, was the foundation of a science distinct from any other science. But a science is not strictly adequate to the thing known when it does not comprehend that thing perfectly, does not comprehend it alone, attains several things under some common aspect, and thus co-ordinates several ideas under the same principle of science and understanding. Such a science is not divided according to the entitative division of things; it comprehends several quiddities under the same specific aspect. Thus, the formal and specific object of a science embraces several physical species; it is not established by congruence with particular real species." Ibid., pp. 560–61.

5. By way of a direction likely to guide the research project, let us note that the pluralist conception of the system of sciences in the Thomist school is logically arranged according to the following phases: (*a*) The distribution of the three orders of abstraction (physical, mathematical, and metaphysical). It is indispensable to use the expression *orders* of abstraction rather than the expression *degrees* of abstraction, which we will have need of later on. (*b*) The theory of mixed sciences, which belongs to an order of abstraction by its matter, but to another order of abstraction by its form. (*c*) The distinction of degrees of terminal abstrac-

tion within an order of abstraction defined by an initial abstraction (John of St. Thomas, *Logica* II, q. 27. a. 1). The examples given by John of St. Thomas are the philosophy of nature and medicine as within the first order of abstraction, arithmetic and geometry within the second order, and metaphysics, logic, and theology within the third order of abstraction. (*d*) After having considered the divisions that result from the diversity of initial abstraction (order of abstraction) and the divisions that result from terminal abstraction (degrees of abstraction), it remains to inquire in what way the division of a science into treatises affects the unity of the science. Take for instance the philosophy of nature. Its order of abstraction is clearly defined by Saint Thomas (*a materia singulari tantum*). John of St. Thomas was aware that the philosophy of nature did not include the entire order of physical abstraction, but corresponded to a degree defined within this order. In fact the only science he knew that was distinct from the philosophy of nature within the first order of abstraction was medicine (ibid., p. 562), a rather troublesome instance inasmuch as medicine is a study of a practical character. Today we know that the first order of abstraction includes, in addition to the philosophy of nature, the entire positive science of nature (at least concerning its matter, for its form might belong to the order of mathematical abstraction). Once it is clearly distinguished from positive science, the philosophy of nature remains divided into a certain number of treatises, and Aristotle's system of physical treatises, as it is explained by Saint Thomas, retains all of its meaning. (Cf. Saint Thomas, *Comm in Phys.* I, lect. 1.) In the first place, there is the treatise on mobile being in general (*Physica*); secondly, there is the treatise on local motion (*De caelo*); thirdly, there is the treatise on substantial change (*De generatione, De meteoribus*); fourthly, there is the treatise on inanimate mixtures (*De mineralibus*); and, finally, the treatise on animated beings (*De anima* and its appendices, *Parva Naturalia*). Let us note in passing that our programs have simplified all of that; we know nothing other than cosmology and psychology. Cosmology includes all those subjects considered by Aristotle in the first four treatises, sometimes even a part of the subjects considered in the fifth treatise (on vegetable life). It would be appropriate to raise the question how well-founded this simplification is. It is a matter, then, of knowing in what way and to what extent this division into treatises affects the unity of the science. Saint Thomas, as far as we know, expressed himself on this subject only in vaguely defined terms. *Comm. in Physics* I, lect. 1; *Comm. in de Sensu,* lect. 1. He states that the different parts of the philosophy of nature are distinguished from each other by the more or less heightened abstraction of their objects. So it is the same principle, namely, the uneven level of abstraction, that determines the division in the system of sciences and the internal division of each science. ("And as different kinds of science are distinguished according to the different ways in which things are separable from matter, so even in the special sciences, and especially in the science of nature, the parts of the science are distinguished according to the mode of abstraction and concretion. And because universals are more separated from matter, therefore in natural science one proceeds from the more universal to the less universal . . ." (*Comm. in de Sensu*, lect. 1). But what is the significance of this internal division? Is it compatible with the specific unity of the science, or must we say that it entails a breach in this unity? If the latter part of the alternative were true, we should say that the philosophy of nature is only a genus and that the ultimate scientific species is constituted by each of the treatises that make up the philosophy of nature. The two greatest commentators on Saint Thomas disagree on this issue. Cajetan declares himself in favor of the pluralist response; he even seems to take it as obvious. Here is what he says, in an incidental way, but with great clarity in the course of a discussion on the definition of pride (*Comm. in II–II*, 162, 1): "Whence by agreement eminence can be articulated in a twofold way as having a different meaning from these notions [namely, presumption, vainglory, and expressed]. On the one hand, as in another matter, as eminent punishment is said to have a different meaning from eminent honor. On the other hand, as eminence in an absolute sense is formally distinct from concrete instances of eminence in such and such a matter. So motion absolutely speaking has another meaning than applied motion, for instance, local motion, alteration, generation and expressed. *And that it is the formal distinction alone that suffices to distinguish specific habitus is evident from the fact that the science of motion taken absolutely which is treated in the books of the Physics, the science of locomotion treated in the book On the Heavens, and the science that treats of motion in regard to form taken absolutely, treated in the book On Generation and Corruption, are specifically distinct.*" John of St. Thomas, on the contrary, holds that the philosophy of nature has the unity of an ultimate scientific species. (*Philosophia naturalis*, I, 1, 2, *Cursus philosophicus* II, p. 17.

6. See "Elimination of Metaphysics," in *Logical Positivism*, ed. A. J. Ayer (Glencoe, Ill.: The Free Press, 1950), pp. 60–81 [ed. note].

5

Some Remarks on the Object of Physical Knowledge

THE EXPRESSION "physical knowledge" is taken here in the sense that ancient philosophers understood the expression *physica, scientia physica, scientia naturalis, philosophia naturalis;* it designates, then, the whole of the sciences, whether philosophic or positive, belonging to the first order of abstraction.

The Thomists say that the object of physical knowledge is mobile or sensible being, material being. Although inquiring minds prefer the term *ens mobile,* Thomist doctrine allows the use of three terms as equivalent: *ens mobile, ens sensible, ens materiale.*[1] We propose a reflection on the equivalence of these terms.

Science has that which is necessary as its object: this principle runs the risk of being insufficiently understood if one limits oneself to the accounts supplied by textbooks and routine schooling. Unless we are mistaken, here is how one proceeds to define science in the ordinary teaching of Aristotelian philosophy.

We take a nominal definition as our starting point: science is a qualitatively perfect kind of knowledge. To be more precise: we say that science concerns a qualitatively perfect knowledge insofar as it is knowledge, that is, provided with theoretical perfection. Science is distinguished by that feature from the *habitus* of practical intellect, art, and prudence, whose perfection pertains to the order of direction and not to the order of knowledge.

In order for theoretical knowledge to be qualitatively perfect, it must be certain above all else. In order for it to be certain, it must deal with an object that cannot be other than it is, a necessary object. But individual existing things are contingent. Science deals with the universal, not because the singular thing as such is recalcitrant to its requirements, but because only in the universal that does it find the necessity it needs to establish its certainty. Existential situations that are the object of historical research are foreign to the scientific object. Finally, perfect knowledge should be explanatory knowledge, a knowledge of things by their causes or by their reasons.

This current description is silent about a very interesting difficulty. Not only is there necessity in the universal. There is necessity in the most contingent realities. As Saint Thomas says, it is contingent that Socrates runs, but it is necessary that he is in motion if he runs. In an historical event that boggles the mind by the mystery of its contingency—Napoleon's defeat at Waterloo—there is an element of necessity on which the certainty of historical knowledge unshakably relies, and that is the necessity of the identity of the event with itself. Since June 18, 1815, the proposition *Napoleon was defeated at Waterloo* is an eternal verity. Moreover, whether it is a matter of universal essences or singular existential situations, is it not the case that the object of the intellect is being? Now, self-identity is the primary law of being, and this law is absolutely necessary. To speak of an intelligible object is to speak of an object necessary in some respect. Whatever is presented to the grasp of the intellect includes some necessary aspect, a certain amount of necessity.

Yet the entire history of the human mind in its effort to construct science shows that qualitatively perfect knowledge is not adapted to any kind of necessity whatsoever, that the kind of necessity involved in a contingent existential situation is insufficient for it, that what it needs is an object implying the higher necessity of universal essences and immutable laws.

In order to show the rationale for this requirement, two basic characteristics of perfect knowledge must be considered together and as interconnected: it is certain and it is explanatory, certain in its explanations, explanatory in its certainties. Let us reiterate that as certain knowledge it requires a necessary object. Now the formulation has a fully determined meaning. Now we understand that science cannot be content with any kind or any degree of necessity whatsoever. It has to have an explanatory necessity. *Now, as we showed earlier, simply historical necessity is not explanatory; only essential necessity would be so.* The scientific object—and it is this which distinguishes it in the class of simply intelligible objects—is necessary with an essential necessity, because it is on this condition alone that it establishes an explanatory knowledge certain in its explanations.

Such a system of essential necessities, or, what amounts to the same thing, a formulation of eternally necessary possibilities, is, then, in absolutely general terms, the object of scientific knowledge.

Let us recall now that a division to be essential should proceed from the viewpoint of the very essence of the divided object. The particularities that essentially divide scientific objects are those that affect them as "systems of essential necessities," as "systems of explanatory necessities," as "formulations of eternally necessary possibilities." There are different and uneven ways for the scientific object to be "a system of essential necessities," "a system of explanatory necessities," "a formulation of eternally necessary possibilities." The different categories of scientific objects are determined by the different and uneven ways in which a scientific object is in essence the systems and formulation mentioned above.

It is precisely here that the idea of a typological abstraction comes in. These systems of explanatory necessities are not the immediate data of experience. They have to be extracted or abstracted from the immediate data of experience by the mind. Now, in every abstraction two terms are at stake: (1) *what* one abstracts, sorts out, or extracts; (2) that *from which* an abstraction is made, that which is left aside, that *from which* one intends to sort out what one wants to sort out. We have sufficiently defined the first of these terms. In regard to the second, we ask the following questions: What is there in the immediate data of experience that conflicts with the characteristics of the scientific object? Of all those aspects which the observable world reveals to experience, which one is opposed directly, properly, formally, diametrically to the characteristic requirements of the scientific object? The answer is obvious. It is by its mobility, by its ability to become what it is not, by its inability to remain what it is, that the thing immediately given in our experience disappointments the expectations of the scientific mind, eager to grasp the *formulations of eternally necessary possibilities.* There is nothing for the scientific mind to do in this river in which one cannot plunge twice.

Little isles of stability appear, however, as soon as the mind passes from consideration of the individual to consideration of the universal. The two operations, called *abstractio formalis* and *abstractio totalis* (let us say, for want of better terms, typological abstraction and extensive abstraction), maintain a special relationship here. They rigorously accompany each other and by the same token supplement each other to the point of appearing to be identical.[2] Let us briefly recall that *abstractio totalis* is the operation by which the mind sorts out a logical whole, a universal whole, the subjective parts being included in its extension. *Abstractio formalis* is the operation by which the mind sorts out a form, a type, that is, a principle of real determination and of intelligibility, from the subjects that support it and owe their determination and their intelligibility to it. We say, then, that, in the first movement of the scientific mind searching for the object that immediate experience denies it, extensive abstraction and typological abstraction simultaneously complete each other. The simple passage from the subjective part to the universal whole, the simple passage from the individual instance to the species and the genus, conveys to the mind a type, a form, an essence, the formulation of a system of eternally necessary possibilities.[3]

Physical knowledge stops the process of abstraction at this point. That is what makes people say sometimes that physical abstraction is merely *abstractio totalis*. It would be more exact to say that the *abstractio formalis* that defines physical knowledge does not go beyond what is obtained when the universal whole is sorted out of the singular realities given in our experience. From the very viewpoint of what constitutes the essence of the scientific object, how is the result of this first *abstractio formalis*—this scientific object immediately, freshly extracted from the waters of observable becoming—to be described?

It is a universal and necessary type. It is contingent that it exists in fact, but it is necessary that it be such and such, and if it exists, with all the constitutive and derived properties.[4] And it is necessary that it be thought of as such and such if it is thought of at all. Extracted from a world in which everything concrete, everything complete is subject to corruption, the universal and necessary type is incorruptible. What difference does it make to science if numerous living species have ceased to be represented on this earth, provided that we have the means to know their specific characteristics? Perhaps future zoologists will add the sad note to the description of the buffalo, the elephant, and the whale: a species that is extinct today. This additional remark will be of a completely historical and practical interest, for there will be nothing to change in the description of the species declared to be extinct.

But the incorruptible type grasped by the mind in the flux of things can exist only in a concrete perishable thing. The mind, in order to pass from a merely represented existence to a real existence, must become involved in a comprehensive whole in addition to the type itself, *a subject carrier of a type*. The whole so composed is susceptible to corruption. The object of physical science is the incorruptible formulation of a corruptible reality. Moreover, if it were thought of as the formulation of a simple thing, this incorruptible formulation would be thought of in an inexact way, misconstrued, and radically falsified. It must be thought of as a formulation of a whole of which it is only a part; it must be thought of as the determined part of a compound thing; it must be thought of in its relation to the other part of the whole in which it is made to exist. It must be thought of as a form of being characterized by the capacity to cease being, *ens mobile simpliciter*.

The adverb *simpliciter*, occasionally added to the two words that make up the current definition of the physical object by Saint Thomas, contributes a useful precision and avoids an ambiguity. There is only God who is immutable in an absolute sense. Let us not conclude, however, that every creature, actual or possible, is included in the sphere of the object of physics, *ens mobile*. Not every being capable of changing in any way whatsoever is the object of physical knowledge, but only the being capable of change in a fundamental way, the being that harbors deep within its constitution a tendency not to be. It is the corruptible being, the being that can neither exist nor be thought of without an essential relation to an inner principle of destruction: *mobile being, absolutely speaking*.

What we call matter is this inner principle of destruction, this partner of the necessary and universal type in the makeup of the physical thing, this subject carrier of the type. I mean indeed matter in the absolute sense of the term, prime matter, absolute matter. Mobile being, the object of physics, is the being in whose substance is included a principle which is itself undetermined and, by reason of its intrinsic lack of determination, capable of casting off its basic determination in order to put on another. The type made to exist with such a principle is affected by destiny in its very typical structure and obliged to be but the formal part of a whole that includes matter. An incorruptible formulation of a corruptible reality, it is affected in its very constitution as formulation by its relation to the matter that it is made to determine.[5] Physical definitions abstract from matter considered as a principle of individuation; without this abstraction, there is no explanatory necessity and no science. But the definitions do not abstract from matter envisaged as a co-principle affecting the universal type in its very constitution as a type. The equivalence between the expressions *ens mobile* and *ens materiale* is therefore justified. Yet the expression *ens mobile* keeps its priority inasmuch as the idea of mobility concerns the essence of the scientific object taken as such in a more direct and absolutely primary way.

The equivalence of the terms *ens mobile* and *ens sensible* expresses a basic aspect of Aristotelianism. The entire philosophy of Aristotle assumes that there is a determinate correspondence between the property of things to appear to the senses—let us say their objective sensibility—and their substantial mobility; that the observable aspects of things constitute a sign, an expression of their mutability and their materiality; and that the mind by referring to the observable aspects of things in their definitions and descriptions knows them in their mutability and their materiality.

Common sense is familiar with the identification of the visible and the perishable. It raises great difficulties, however, as does everything that touches the mystery of sensation. The first thing to do in order to be aware of these difficulties and prepare a solution to them no doubt consists in realizing the instability of the world in a more direct and more vital way than we usually do. In our daily lives, we have a tendency to underestimate the breadth and the depth of this instability through the effect of illusions quite easily explained by our capability as intelligent and practical people.

Let us refer to the Aristotelian division of sensible objects. In addition to the objects that are inherently sensible, that is, objects that exercise an action on the sense organ and on the sensory power, we claim to see, hear, and touch a lot of objects that are not at all inherently sensible: substances, relations, works of art, useful or harmful tendencies, and so forth. It is obvious that no sensory power can know a relation. Yet we do not believe we are lying when we say: I hear my father's voice. The fact is that numerous objects pertaining by themselves to intellectual or instinctive cognition are accidentally constituted as sensible thanks to their constant connection with the inherently sensible in the concrete organization of our perceptions. I shall call "pure sensation" the act of external sensation whose object is the inherently sensible, and I shall call "perception," or "concrete perception," the complex act in which every knowing power plays or can play a role, and whose object is a whole including the inherently sensible and all the accidentally sensibles that form a psychological unit with the inherently sensible.

If the world of our sense experience seems to present more stability than it really does, it is because we pay less attention to the inherently sensible, to the objects of pure sensation, than we do to the accidentally sensible involved in our concrete perceptions. The environment of our daily experience is pervaded with intelligible and practical meanings. We completely turn our attention to these meanings because we are intelligent and in a hurry. There would have to be exceptional circumstances in order for our attention to the intelligible and the practical to slacken to the point of allowing us to notice, and not without astonishment and discomfort, that our sensory environment is a current that never stops and never repeats itself. When a poet, inverting a phrase in the Gospel, exclaims "love what you shall never see twice," he utters the exact formulation of the *conversio ad sensibilia.*

The inherently sensible, the object of pure sensation, is itself a complex reality. In terms of Aristotle's analysis, it includes the proper sensible, that is, the sensible quality that can be attained by only one sensory power, and the common sensible which is a determination of the sensible quality such as form, size, number, states of motion and rest, that different sensory powers can attain by means of the proper sensibles. Thus, we could count a collection of marbles either by looking at them, or by touching them, or by hearing them fall one by one. But one must insist on one point: the common sensible appears to the senses only as a modification of a sensible quality; hence, the sensory world is, above all, a world of sensible qualities.

Now, if we return to the general principles of the theory of quality, the mutability of the sensory world immediately takes on the characteristics of a paradox loaded with meaning.

The best way of explaining the Aristotelian definition of quantity and of quality no doubt consists in showing what happens when we explore a quantitative whole, on the one hand, and a whole made up of different qualities, on the other. If we consider this work table from a quantitative perspective, it is a rectangular figure. I run my hand along or glance from the part located in the back to the part located in front, from the part located at the left to the part located on the right. As long as I limit myself to a purely quantitative viewpoint passing from one part of a quantitative whole to another part, I pass from an object of the same kind to an object of the same kind. The parts of a quantitative whole are *homogeneous* and distinguished from each other only by their extrapolation (*quantitas est accidens tribuens subjecto habere partes extra partes quoad se*). Let us now consider a whole made up of qualities, such as the entirety of the dispositions of the soul. When I pass from science to prudence, from justice to moderation, I pass from an object of a certain kind to an object of a different kind, from the same to the other. Or, again, let us glance once more at this table, but this time viewing it as a surface bearing qualities. Here is a dark line; there, a vividly illuminated area. Here is a black spot; there, a whitish line. The parts of the qualitative whole are mutually *heterogeneous*. *Quality is what makes a thing to be such and such.*

This opposition between the homogeneity of the parts of the quantitative whole and the heterogeneity of the parts of a qualitative whole reveals the analogy existing between quantity and matter, on the one hand, and between quality and substantial form, on the other. The first analogy may be expressed by stating that if prime matter could exist in a separated state and could be observed in that state, it would present a sight of indefinitely monotonous indetermination to the spectator. The second analogy will be expressed by stating that the substantial form is what makes a thing to be *this or that,* even as quality is what makes a thing to be *such and such.* But it is not just a question of a resemblance between quality and form, between quantity and matter. There is a relationship of real derivation. Quantity really proceeds from matter and quality really proceeds from form. That is why, in our effort to know the proper constituents of things, what specifies and distinguishes them from what they are not, we spontaneously turn our attention to their qualities and the operations that reveal their qualitative determinations.

It is here that the paradox of the mutability of sensible qualities appears. The form, principle of determination, of distinction, of heterogeneity, measure of being, plan, idea, rationale, is the principle of stability, of constancy, and of permanence. It is the form that confers self-identity at the same time as distinction forms all that is not the self. Matter, on the contrary, principle of indetermination of passive indifference, a gap opened in the structure of being through which new forms can be introduced by destroying the presently established structure, is the very principle of inconstancy, of fragility, of precariousness—in a word, of mobility. If quality proceeds from the form, it is to be expected that every quality possesses the characteristics of the form, the properties of stability, constancy, and permanence. This expectation is dashed by the fluidity of sensible qualities.

And so we are led to understand that sensible qualities express not only the formal constituents of things, considered precisely insofar as they are formal constituents, but also the mode of being imposed on them by their nature as forms destined to exist in matter, by their association with a principle of corruptibility, and by their characteristics as parts of corruptible wholes. Taken as qualities, sensible qualities can be said to express the form of being that they reveal; and their mutability expresses the conditions that affect the form by virtue of the essential relation it maintains with matter.

Let us return now to the viewpoint defined by the notion of scientific knowledge. It must be said that all scientific knowledge has *a formulation of eternally necessary possibilities* as its object. In order to arrive at this formulation, there must be an abstraction from the contingency of existential situations (for the element of necessity that subsists in this contingency cannot be the principle of explanation). *Abstractio totalis,* the process by which the mind sorts out the universal from the multiplicity of individual instances, implies *abstractio formalis,* the extraction of a type independent of any particular realization in individual existence. The process of abstraction in physical knowledge stops here.

The scientific object so obtained, as a scientific object and as the *necessary law of possibility,* is characterized by the following property: the necessary laws of possibility that constitute the scientific object imply, if the object exists, the requirements of existing in a precarious and impermanent way, that is, as a mobile being, or, what amounts to the same thing, in a state of association with a radical principle of mobility, that is, as a *material being*.

But this radical principle of mobility is disclosed to our experience by the special condition that it imposes on the observable qualities of mobile things. By the very fact that it associates distinction and inconstancy, sensible quality is the proper expression of the ontological type constrained by virtue of its constitution as a system of necessary possibilities to be but the incorruptible formulation of a corruptible reality. In a definition or a description, to mention the sensible qualities of an object, to know it as a *sensible being*, is thus to know this object as mobile being. It is to have a properly physical knowledge of this object.

The expression *materia sensibilis*, which seems so obscure to beginners—how could the senses perceive the matter of things? *nec quid nec quantum nec quale*, and so forth (neither a quiddity or substance nor quantity, quality, or any of the other Aristotelian catagories)—is now quite clear. It simply designates the sensible qualities of things but also usefully reminds us that the ontological types being expressed by sensible qualities owe this essentially fluid mode of expression to the substantial principle of impermanency without which they could neither exist nor be thought about.

Notes

1. The expression *ens materiale* is unusual. We use it as shorthand for the formulation *scientia naturalis abstrahit a materia individuali, non autem a materia communi*.

2. Cf. Saint Thomas, *The Division and Methods of the Sciences*, q. 5, aa. 1 and 2, trans. Armand Maurer (Toronto: Pontifical Institute of Mediaeval Studies, 1986), pp. 9–31; Cajetan, *Commentary on De ente et essentia*, prooemium; Cajetan, *Commentary on the Summa Theologica*, I, 5, 3 ad 4; John of Saint Thomas, *Cursus Theologiae*, I, disp. 6, a. 2; Solesmes edition, I, p. 533; Maritain, *Philosophy of Nature*, pp. 15–26 [note updated, ed.].

3. To eliminate the crude simplifications that scholastic adages risk encouraging, let us mention that the twofold operation described above

is complex, difficult, progressive, always imperfect in some connection. Sorting out a universal whole from its subjective parts is a task that often calls for centuries of speculation. To be convinced of this, it suffices to remark that when one verifies current definitions of certain familiar realities one often notices that they sin against the rule *toti et soli definito*. If the definition of a universal type excludes certain particular cases in which the type is realized (contrary to the rule *toti definito*) or includes particular cases in which the type is not realized (contrary to the rule *soli definito*), it is obvious that the universal whole has been clumsily extracted from its subjective parts and the undertaking has to start over again.

4. We observe that the truth of these propositions is independent of the problem of the boundaries of the specific unity such as it is stated by Charles De Koninck in his article, "Réflexions sur le problème de l'indéterminsme," *Revue Thomiste*, 43 (1937), 227–52, 393–409. De Koninck takes the proposition "natural infrahuman species are only realized in sub-species" (p. 234) as established, and also that the sub-species "are contingent in their very structure" (p. 236). The human being, the animal, the plant, and inorganic substances would be the only "very certain and definable species" (p. 234). Thus, in De Koninck's system, natural types that display the characteristics of absolute determination, which are those of the scientific object, would coincide with the realms that make up the primary divisions of nature and would involve no other diversity. The author believes he is in a position to demonstrate that this is necessarily so.

If the views of De Koninck on the strict coincidence of specific determinations and the realms below human nature are not accepted, if, in other words it is accepted that there are many "certain and definable species" within the inorganic realm, within the plant kingdom, and within the animal kingdom, the uncontested impossibility of producing evidence for the necessity of specific types would stem solely from the imperfection of our means of knowing.

5. This demand of matter introduces an irreducible obscurity into the form itself. One cannot have a distinct idea of it as a cosmic form independent of the idea of the composite and the matter that enters into the idea of the composite is in no way determined without signifying as well determinability in regard to an infinite number of other forms." Ibid., 233.

6
Knowledge of the Soul

NO ONE DENIES that the study of psychology presents us with a scene of awful disorder. The evil, an old one, has been pointed out innumerable times, and has continued to get worse. Among the most apparent symptoms and factors of this disorder are the lack of a strictly determined language, the lack of general agreement about allocating the subject matter and problems, and the disconcerting proliferation of theories. But one must go beyond these well-known facts in order to find the most radical symptom and the most effective factor of a confusion that has strangely persisted. What is most bothersome in the state of psychological research is the general uncertainty about the epistemological nature of psychology. The practitioner of this science has a hard time knowing the kind of science he practices. From one professor to another, from one book to another, the point of view, principles, methods, and affiliations change, and yet the course or book title indicates that it always concerns psychology. It does happen that a psychology course might be especially devoted to theories with a metaphysical aspect on the self and the non-self, the subject and the object, the *a priori* and the empirical, and so forth. It also happens that in the course one speaks especially about the salivation response in the dog and the croaking reflex in the frog. Sometimes the atmosphere of the course is industrial, commercial and medical; it concerns adaptation and performance. Sometimes, though rather rarely, the atmosphere is ethical and literary; the sentences are more polished, and novelists, playwrights, and essayists provide a contribution of the first importance. It also happens that all these types of psychological investigation are combined in the same course or the same book.

The theoretical and practical drawbacks of this confusion are obvious. We will grasp their seriousness better if we manage to identify each of the forms of knowledge arbitrarily confused under the name of psychology. An author whose name eludes me proposed some years ago that we speak of psychology in the plural. Generally this suggestion has been taken as an ironic exaggeration; we take it very seriously. It is perfectly clear that the knowledge of the soul is capable of different forms. It is by no means obvious that this diversity can be reduced to the unity of a single science. Maybe there are several sciences of the soul which it would be wrong to designate, without being much more precise, with the common name of psychology. It is worthwhile testing this hypothesis.

The word "psychology" in ordinary usage is most often related to a kind of knowledge seldom taught in schools, and it is possible to possess it in an eminent degree without ever having read a book in psychology. Here are two doctors equally competent in their art. I prefer one of them because in addition to his art he has an ability that the other lacks. If he feels obliged to acquaint me with a disquieting diagnosis and to prescribe a painful treatment for me, he will know how to choose the right words and attitudes to inspire confidence and courage in me, and at the very moment in which he passes on the bad news, he will help me find the strength to put up with it. The other doctor is not a bad man, but he has a way of presenting things that saps the morale of his patients. He does all he can to comfort them; he succeeds only in depressing them. We say that the first is a good psychologist and that the second lacks psychology. In the same sense we speak of the psychological qualities of a captain of industry, a businessman, or a statesman. Again it is in the same sense that we attribute an enormous psychological insight to certain novelists, historians, playwrights, and to all the great moralists. When the man of action wants to improve his knowledge of men by reading, it would hardly occur to him to read a treatise in psychology designed for college courses. Nor will he read Aristotle's *Treatise on the Soul,* or any of the treatises on the soul written by philosophers following the Aristotelian model. He reads Shakespeare, Balzac, Dostoevsky, Saint Augustine, and Pascal.

The man of action by the very fact that his profession requires him to carry out numerous intense relationships with his colleagues finds himself at the forefront of the kind of knowledge we are trying to describe. The writer, the moralist, maintains a certain distance from men; solitude constitutes part of his vocation. Because of this distance and solitude, it is likely that the psychology expressed in his writings lends itself to a mixture of epistemological types. Therefore it is convenient to observe in the first instance the characteristics of this psychology in the ordinary sense, this everyday psychology of which little is said in the schools of psychology, in the mind of the man of action.

1. The psychology of the man of action is a practical discipline. Behind different and changing disguises, it helps him to recognize real intentions, temperamental data, habits, and motivational priorities. It allows him to predict the reactions of a man in particular circumstances. It allows the determination of the means to employ in order to ascertain that another's conduct serves my purposes. The choice of a partner presents great psychological difficulties for the head of a business. A young man comes to offer his services; his previous employers, people whom one can trust, attest to his honesty. We are reasonably assured that he has never stolen anything. But for that matter he has never had the opportunity to steal an important amount of money. The question, then, is to know if his will, strong enough to resist the temptation to steal small sums, is strong enough to fend off the temptation to steal a lot of money. Let us suppose that observation uncovers the signs of a weak will. Is it a hopeless case? Maybe not. In the candidate's conversation and in his past history we detect a lively sense of honor. Moreover, if his qualifications are satisfactory, it is worthwhile running a risk; let's trust him. By a generous act of trust, let us intensely stimulate the sense of honor which can still protect him against the weakness of his will. Events confirm our expectations. Under the influence of a sense of honor stimulated by trust, the young man is strengthened in his habits of honesty. If the head of the business has to his credit a certain number of these kinds of success, it will be said that he is a good psychologist.

No doubt it is necessary to have many kinds of knowledge in order to obtain frequent good results in dealing with such problems. It is necessary to have observed a lot, retained a lot, compared a lot, assembled and maintained by repeating successful interpretations a flexible and complex system of interpretive associations. But of all these kinds of knowledge, none has its end in itself; each has as its end an action to be directed.

2. The psychology employed by the man of action sticks to considerations of the whole more than to an analysis of isolated

factors. It is not a matter of knowing what a man's imagination or memory or intellect will do in such and such circumstances, but rather of knowing what the conduct of the man will be if he behaves as friend or foe, as a submissive fellow or as a rebel, if he will take the initiatives at an opportune moment or will be content to bow to events. We lack an adequate term to characterize a knowledge which not only does not seek but even avoids the isolation of factors; we will understand the nature of this knowledge well only when we have designated it with an adequate term. In spite of several precedents, let us not speak of a *total* knowledge; this expression suggests the idea of an exhaustive kind of knowledge. Nor do we speak of synthetic knowledge; a synthesis is the act of constructing a whole by reuniting its elements. Now, it is not a question of constructing a whole but of grasping it. For want of a better, we propose the expression *totalistic* knowledge.

A practical psychology in the sense that has just been described is forced to adopt a totalistic viewpoint. And conversely, any totalistic psychology is forced to remain a practical psychology. In order to account for this, it is sufficient to consider that any theoretical science, by the very fact that it proposes the perfection of knowledge as such, seeks an explanation, for perfect knowledge is explanatory knowledge. Now, there is no explanation without a preliminary analysis that takes apart contingent connections and sorts out relations of essential causality which alone are explanatory. In the example mentioned earlier, theoretical knowledge would recognize at least three factors, each of which constitutes an intelligible tendency to produce certain effects: habits of honesty, tendency to produce honest acts; weakness of will, tendency to make dishonest acts possible; a sense of honor, tending to make honesty prevail every time its opposite would involve a feeling of dishonor. These isolated causal relations are intelligible and explanatory; there is no way in which they allow us to foresee the actual conduct of the person being considered. In order to foresee conduct which will be the result of a contingent—therefore unintelligible—association of a plurality of factors, one must consider these factors all together, to see how they fit together in the concreteness of their contingent association; but then one gives up on explanation, and the viewpoint adopted is justified only by the requirements of action.

3. The knowledge of man can abstract from free choice pro-

vided it proceed analytically. Under this condition it is concerned with determined natures, with possibilities defined by the nature of things, with natural determinisms such as the law of sensation, of intellect, and of the will itself. But once the knowledge of man adopts the totalistic viewpoint, free choice takes its place at the very center of these considerations. No doubt there are a great number of actions in the daily conduct of all men which are not immediately regulated by free choice; but the least that one can say is that in the final analysis most of them depend on free decisions.[1] My schedule shows that I must teach a course on Monday at eight o'clock in the morning; it is possible that I act in a determined way when on Sunday evening I set the dial of my alarm clock at 6:45; that at the sound of the alarm I get out of bed without deliberation and freedom; that all the movements that take me to the bus and the classroom are governed by quasi-instinctive judgments from which freedom is absent. But I have performed an act of freedom when I decided to be a professor rather than a lawyer or a doctor; I performed an act of freedom when I decided to follow an honest profession rather than to live a life of laziness and catch as catch can. These abnormal cases aside, free causality is what is deepest, most decisive, and most formal in the causal system from which human conduct results. To ignore the qualifications of this free will, which, in the final analysis, makes the conduct of a man what it is to oblige oneself not to know what contributes more radically than any other factor to make this conduct be what it is. Practical and totalistic psychology is *moral* psychology; that is its correct name. In order to foresee the reactions of a man in particular circumstances, it is no doubt relevant to know if he has good vision or if he is myopic, if he is attentive or distracted, if he has a memory for proper names or if he lacks this form of memory, if he is sensitive or cold, subject or not to irrational fears, if he has a quick or slow mind, if he is very intelligent or only middling. But it is still more important to know if the deepest intentions of this man are good or evil, if he is inclined to tell lies or the truth, if he has a sense of duty, of keeping his word, of justice, the respect for each person's dignity, if he is open to gratitude and compassion. For, ultimately, whatever may be the sharpness of his intellect and the trustworthiness of his memory, the steadiness of his attention and the healthiness of his emotional life, what is important for me to know above everything else to foresee his actual conduct is the use he makes of his powers. From these very elementary remarks it follows that practical and totalistic psychology cannot abstract from moral good and evil in any way.

In order to recognize the distinctive characteristics of moral psychology, we have to consider this discipline in the state in which it is most clearly distinguished from other forms of psychological knowledge, that is, in the state it assumes with the man of action, in the immediate grip of action. In this state it is a knowing power completely directed toward the individual case. All that the man of action has in view is to know this or that particular man. Once he steps back and begins to express his thoughts on man and on different kinds of men that he has come to know, he no longer speaks as a man of action, and his knowledge of moral man assumes a new state characterized by the search for universal types. What we appreciate above all else in the novelists, poets, and playwrights whom we call great psychologists is the ability to embody a universal type in an individual existence, in whom all the—strongly individualized—traits cause the universal type to shine forth. Hamlet and Rastignac are concrete universals.[2]

With the moralist, moral psychology is extracted from the particular features of individual existence and attains the state of abstract universality. For instance, let us mention the celebrated passage of Pascal on diversion; there it concerns a trait of the physiognomy of universal man described in universal terms. On the other hand, one should note that the psychology of moralists is often stated in the form of disconnected jottings. The daily record which follows the divisions of the calendar, isolated thoughts, and fragments are its favorite forms of expression. When the psychology of moralists is organized, its principles of organization are generally rhetorical, aesthetic, and literary, rather than scientific. This is an occasion to wonder if moral psychology, in addition to the state it assumes in the man of action (knowledge of the singular), in the creator of moral types (knowledge of the universal involved in a single existence, real or fictitious), in the moralist (a non-systematic knowledge of the abstract universal), is capable of a state of scientific systematization. In a word, would it be possible to compose a treatise in moral psychology? We see no reason to believe that the difficulties in such an endeavor cannot be overcome.

Moreover, we could mention many attempts at a scientific systematization of facts pertaining to moral psychology. Most bear the name of the discipline called the science of character, characterology. It is appropriate here to ask oneself about the relations between moral psychology and characterology. These disciplines do not coincide as far as content is concerned, for moral psychology encompasses a large number of facts foreign to the object of the science of character. Perhaps we should also say that they do not coincide formally insofar as characterology prefers to fasten on what is determined in the constitution of characters and tends to abstract from free causality. But it is by no means certain that this abstraction is legitimate. If one had seen in this abstraction only the effect of a deterministic epistemology, characterology should be conceived as a part and function of moral psychology.

There is nothing astonishing about the fact that the man of action generally shows no interest in the problems studied in the psychology of the traditional school; they are theoretical problems, not his problems.[3]

Any theoretical psychology, to the extent to which it shares the ideal of theoretical science, has characteristics opposite to those of moral psychology. It seeks explanatory laws rather than formulations of foresight. In its search for explanation, it boldly practices the breaking down of wholes presented by experience; it considers free choice only as a property necessarily resulting from a certain natural constitution; it abstracts completely from the way in which man makes use of his powers. Should one say that it is useless? Like any theoretical science, it is essentially constructed in order to procure an advantage higher than any utility, the knowledge of truth. Moreover it furnishes much information of great practical value. But, according to the universal law of theoretical thought, the information of practical interest is characteristically the effect of superabundance, and surely the best way to obtain this information is not to seek it.

The terminology of scholastic programs and textbooks appears extremely uncertain in regard to the scientific unity of theoretical psychology. Generally it is admitted that there are two theoretical sciences of the soul. One of them is akin to philosophy; the other, to positive science. To teach the first we prefer a man of philosophical background who has read Plato, Aristotle, Lucretius, Saint Thomas, Descartes, Leibniz, Kant, J. S. Mill, and Renouvier. To teach the second, we prefer a man who has spent some time in laboratories and psychiatric hospitals. Current ideas hardly admit of a more precise division. The terminology is changing and confused. Metaphysics, metaphysical psychology, rational psychology, speculative psychology, philosophy of mind are the expressions most often used to designate this so badly defined science of the soul, regarding which there is a vague agreement to acknowledge it as especially a matter for philosophers. Psychology (without a modifier), experimental psychology, positive psychology, scientific psychology are so many expressions serving to designate this science of the soul to which we attribute in a confused way the honorific title of positive science, without bothering too much as to what this last expression exactly means.

Concerning the theoretical form of the knowledge of the soul, the problem of psychology is only a particular case, but a particularly difficult one, of the relationship between positive science and the philosophy of nature. Recent transformations occurring in the structure of physics have reminded us that the oldest established positive sciences have not yet succeeded in achieving the conquest of their autonomy. A positive psychology exists; it is recently established. Numerous researchers have enthusiastically pursued the ideal of a fully positive psychology, as independent of philosophy as chemistry and biology are; some of them have become discouraged. During the last fifty years, many psychologists have come to admit, with or without sadness, that a state of confused association with philosophy could well be the permanent condition of so-called positive psychology.[4]

This resignation appears to us to be an unhealthy thing, as is every attitude destined to perpetuate disorder in the system of our knowledge. No doubt, there is no essential reason prohibiting the science of the soul from putting on an entirely positive form, but different accidental reasons abundantly explain the failures suffered by positive psychologies in their endeavors at autonomous system-building, without implying that these failures are definitive. Let us briefly indicate some of these reasons.

1. Unlike what occurs in physics, chemistry, and biology, psychological facts are for the most part familiar to common sense. Now, common sense, when it does not become absorbed in prac-

tical preoccupations or deceived by the figments of the imagination, is much more inclined toward a philosophical than a positive interpretation. It follows that the elaboration of a positive concept in psychology often consists in the recasting of a philosophical concept. Such a recasting requires an extremely vigilant critical sense and is effected only through trial and error. In the case of protracted failure, the autonomy of the positive synthesis becomes compromised by the violent and, moreover, generally unconscious incorporation of a philosophical concept.

2. It has often been remarked that positive science is the work of a sagacious reason, capable of controlling the ardor of its natural penchant for the being of things. This remark is particularly relevant for the knowledge of the soul. The rational appetite for the being of things makes itself felt with a particular vehemence when the thing to be known is the very principle of our life and the subject of our destiny. Unless he is a pure empiricist gone astray in theoretical science, the positive psychologist must fight without letup against the ontological enthusiasm that threatens at any moment to change the epistemological nature of his interpretations.

3. In the investigation of sub-human nature, few facts lend themselves to an ontological interpretation. The frequent and evident failures of attempts at a philosophical explanation effectively protect positive reason against the interventions of philosophy. In things of the soul, on the contrary, especially when higher functions are at issue, a considerable proportion of facts entail an ontological interpretation. The philosophical mind has more occasions to become conscious of its possibilities and fewer occasions to recognize its limitations.

4. Finally, it is appropriate to remark that the success of a scientific systematization is conditioned by the possibility of exploring the set of facts whose systematic expression is at issue in an ongoing way. If the positive datum presents numerous gaps, it is very difficult, and maybe impossible, to achieve a satisfactory systematization through the use of positive principles alone. In case of failure, it is necessary either to give up the advantages of the systematic form or to borrow the principles of systematization from philosophy.

Positive psychology often borrows its principles of systematization for the simple reason that the positive datum with which it deals presents important gaps. The familiar objections of Auguste

Comte against the very possibility of a positive science of psychic facts are far from being groundless; they are only exaggerated. They do not demonstrate that a positive psychology is impossible, but they are a timely reminder that the organization of this science poses exceptional difficulties. More than in any other science, observation in psychology runs the risk of self-destruction by altering the process under observation. As soon as I begin to observe my emotions, they stop being what they would have been had I not observed them. If I question a sick person suffering from an anxiety neurosis, what his statements make known to me is less the consciousness of an anxious person than the consciousness of an anxious person who knows that he is being observed by a psychologist and adapts to this particular situation. In a large number of cases, moral feelings set up a lively resistance to stating the truth. In regard to certain very important aspects of the emotional life, this resistance of moral feeling is only actually overcome by the patient in whom the desire to be cured overwhelms the inhibitions of modesty and pride. The observation of a man in good health is missing from the file.

The cleverness of experimentalists has produced countless techniques aimed at getting around all these difficulties and many others; the results of their research are still far from presenting the picture of a continuous and easily arranged whole. Yet, whatever may be the data of experience which are still missing, there is room for thinking that a better interpretation of the epistemological situation would allow the realization of great progress in the organization of positive psychology. As an effect of mental habits generated by specialization, different research techniques become hardened and isolated to the point that many minds give up seeing them integrated in an organic whole. Introspective psychology, physiological psychology, animal psychology, and abnormal psychology are spoken of as if they were quite distinct sciences. There is a consensus that they only maintain occasional relations with each other and pursue their development in a dispersed state. If we knew better what we were talking about when we talk about positive science, we would understand without any difficulty that these different "psychologies" are in reality only different paths of research contained within the unity of positive knowledge and aimed at realizing in a rough way the continuous exploration of psychic facts by the contribution of complementary findings. The discontinuity of positive data that we have just described is only partially due to the intrinsic difficulties of psychological observation; to a large extent it results from the inadequacy of our epistemological ideas.

Thus, far from wishing that psychology forgo constituting itself an autonomous positive science and agree to let itself be organized by philosophy, as many philosophers are inclined to do, we wish that a growth in epistemological awareness would make the too-little-known possibilities of positive systematization apparent to the eyes of psychologists, along with an objective unity annoyingly disguised by the diversity of methods.

This progress in epistemological consciousness could be facilitated by a reform of vocabulary; it will have little chance of being produced as long as we use all kinds of arbitrary expressions to designate the two great forms of theoretical psychology.

In regard to that science of the soul which is part of philosophy, or, more precisely, a part of the philosophy of nature, the simplest expression is also the most adequate: we shall call it *philosophical psychology*. Not only is the expression "metaphysical psychology" inexact, it is contradictory, for the soul, *psyché,* is not beyond nature but in nature; it belongs not to the metaphysical but to the physical world. It is unquestionably possible to carry on a metaphysical study of knowledge, appetition, intellect, and will, but to take this metaphysical study for a psychological study is simply to ignore the peculiarities that affect the perfections of being when they are involved in the world of mobility; it is to ignore the animating role of the soul.[5] The expression "rational psychology" is equivocal and disturbing; it suggests a bad sort of apriorism. The expression "speculative psychology" has a definite meaning only in contrast with the expression "practical psychology"; hence, it implies that any non-philosophical science of the soul is a practical science, and that is false. The expression "philosophy of mind" reduces the realm of philosophical psychology in an arbitrary way by suggesting that only the rational part of the soul is susceptible to philosophical analysis.

Concerning the positive science of psychic facts, the only correct expression is *positive psychology*. The expression "experimental psychology" is too narrow and, strictly speaking, designates only research carried out by means of experiments, excluding research carried out by observing non-induced phenomena. The expression "general psychology" is devoid of any exact meaning; the expression "scientific psychology" inconveniently suggests that philosophical psychology has nothing scientific about it and belongs to literature, a fantastic conception that many positive psychologists are only too willing to accept.

Most of the positive sciences generate a technology. Let us call *applied psychology* the technique or set of techniques that derive from positive psychology. So, in addition to theoretical psychologies, two practical psychologies exist: moral psychology and applied psychology.[6] The big question is to understand rightly in what way they are to be distinguished, under the same heading of practical sciences.

Each of them sets out to know man in order to act upon him, to foresee his conduct in order to direct it. Now, two causal systems exist in the human soul: the system of determined causality and the system of free causality. Each of these systems provides a distinctive mode of interpretation, foresight, and influence. It is possible to understand a man's free acts by relating them to the dispositions of his free will; it is possible to predict his free reactions with varying degrees of probability; it is possible to affect his free behavior by modifying the dispositions of his freedom. It is possible to understand his determined reactions by relating them to their determining causes, to foresee them sometimes with a high degree of confidence, and to modify them by transforming their determining causes. The first mode of interpretation, foresight, and influence is that of moral psychology; the second is that of applied psychology.

If we deny the existence of free choice, we find ourselves obliged to transfer the functions of moral psychology to applied psychology, and man is surrendered to a technique whose primary task is to achieve the suppression of freedom by any means whether crude or subtle. In fact, as long as freedom refuses to be suppressed, the technical knowledge of human action will suffer numerous setbacks in its imperialistic endeavors. Countless minds are obsessed by the ambition of a technical knowledge extended to all the spheres of human action and absorbing the knowledge of moral man for its own benefit. This constitutes an exceptionally serious threat. It is a threat all the more formidable since it is often hard to draw a line of demarcation between the realm of determined causality and that of free causality, between the possibilities of technical knowledge and those of moral knowledge. We will try to show by several clear examples how it is possible to draw distinctions between these two realms in typical cases. A distinction based on the certitude of typical cases may still guide thought in the obscurity of confusing situations.

1. A witness states before a court that he has seen a woman wearing a red dress in an unlighted alley at 6 o'clock at night. It is sufficient to apply a simple law of positive psychology to know that the testimony is substantially false; the human eye cannot distinguish red from black in the dark. But the judge needs to know something else. It is necessary that he know whether the witness is an honest person fooled by his imagination or if he is trying to fool the court. To verify the sincerity of a witness is a problem in moral psychology.

2. A public transportation company is involved in hiring drivers to serve a particularly dangerous route. The first thing to do in examining the candidates is to test the state of their sensory and sensory-motor functions, which are determined functions. Several rules of applied psychology will allow one to recognize the candidates who should be considered unfit regardless of their good will: the color-blind, the myopic, subjects whose reaction time varies greatly or who show themselves incapable of sustained attention will be eliminated without any more ado. As to the subjects considered fit, one will not entrust them with the driving of a bus without being assured that they possess certain dispositions such as temperance, discipline, habits of regularity, and a sense of responsibility. This second part of the investigation is a matter of moral psychology.

3. In the preceding examples, the respective roles of applied psychology and moral psychology are so sharply distinct that they can be conveniently separated. It is not necessary that the technician assigned to measure reaction time be a psychologist in the ordinary sense of the word, an expert in the human heart. It is enough that he knows how to operate an apparatus and make a calculation. On the contrary, sometimes the problems of applied psychology and those of moral psychology are so mixed together that their borders are practically indiscernible: it is what occurs in the exercise of psychotherapy in all its forms, in pedagogical activities, in research on the most favorable conditions for industrial work output.

A psychologist in the factory will discover, for example, that a certain change in the lighting arrangement is accompanied by an increase in output. The interpretation of this fact may be the province either of applied psychology or of moral psychology or of both disciplines. It is possible that the new lighting system operates by way of determined causality, by making perception easier and lessening fatigue; it is equally possible that it operates by way of moral causality, by indicating to the worker that he is the object of attentive consideration and thus stimulating his self-esteem and good will. No doubt it is very interesting to recognize the role due to each causal system in the production of an observed result. But what is quite certain is that if one is preoccupied with actual performance, it cannot be a question of treating separately the determined factors attributable to applied psychology, and the influences which apply to free will.

The progress achieved by applied psychology in recent generations has conferred a new attraction to the most daring of scientistic ambitions: thanks to the positive science of the human soul—or if one prefers, of human behavior—the utopia of mankind's exercising a control over itself analogous to that which it exercises with an ever-increasing success over irrational nature.[7] This utopia requires the suppression of free choice; that does not mean that it is completely unrealizable. To the extent to which it is possible to suppress man's inner freedom, applied psychology promises potentates a power that no industrial science would have given them; souls themselves are placed at their mercy.

In fact, that is what the tragic experiences of our time have taught us: psychological techniques, which can be exercised only on determined causes, have the power to create the subject on which they want to practice. In fact the tyrannies of the past had only physical means as instruments of constraint; today's tyrannies have *psychic constraint* at their disposal.

The concept of constraint is usually associated with the idea of physical force, and so the expression *psychic constraint* may seem to be a contradiction. Yet hypnotic and post-hypnotic suggestion provide well-known instances of psychic influences which are not at all processes of persuasion but, indeed, processes of constraint. What should we say about propaganda? A moderate kind of propaganda is a persuasive process; it is a moral influence tending to generate certain dispositions in the person's free will. An intensive kind of propaganda, especially if it is not checked by any counter-propaganda, is a process of psychic constraint, comparable to hypnotism but capable of gaining the submission of countless wills in widely different fields of activity. Totalitarian states have used armies of psychologists to achieve this breakdown of inner defenses without which their social and military endeavors would have been impossible.

By considering what has been accomplished by totalitarian states, we can form a rather exact idea of what mankind claiming to guarantee control over its destiny by means of a technology of human phenomena would be like. In order to ensure the triumph of this *technology extended to man,* one would have to have an absolute and irresistible power, capable of suppressing any dissenting opinion. In the silence of an unbounded despotism, a gigantic scientific mechanism of psychic constraint would complete the annihilation of inner freedoms and would endeavor to remake human desires according to a model dictated to psychologists by their employers. Let us be aware that this utopia has already received important initiatives toward its realization.

We beg the reader not to see in these reflections the sign of any ill feeling whatever in regard to applied psychology. We are not among those moralists who believe that nervous disorders are healed with edifying discourses. We are in no way inclined to think that preaching virtue makes the task of procuring the great benefits of mental health and a successful adaptation to the natural and technical environment for men superfluous. On the contrary, we believe that the healthiness of psychic functions should be counted among the number of conditions that most effectively promote the development of the virtues, and that whoever is interested in real morality must wish that the possibilities of applied psychology in all its forms be thoroughly exploited.[8] It is just a matter of respecting real differences whose distinction concerns the salvation of persons and societies in a direct way.

It is easy to lay all the responsibility for the evil on wicked pragmatists and materialists who try, with some success, to utilize the positive science of the soul and its applications for the victory of their conception of human destiny. But these sinister attempts gain a valuable advantage from the confusion of ideas about the knowledge of the soul. And so there is a need to return to principles in order to introduce clarity into the epistemological situation of the psychological sciences and to go beyond doctrinal chaos.

In ending, we would like to call attention to the role of scholastic programs in this indispensable work of clarification. Most students spontaneously believe that the pedagogical division of educational subject matter coincides with the real division of the sciences. In fact, it would be natural and desirable were it so and if the order of scholastic programs helped to spread exact ideas on the order of our knowledge. In regard to the science of the soul—and perhaps certain other sciences—school programs unfortunately feel the effects of contingent factors which have marked the development of discoveries, doctrines, and publications. No doubt it was inevitable. But the time has come to conceptualize a reorganization of the teaching of the psychological sciences according to the data of epistemological reflection.

1. First of all, we should especially convince ourselves that questions of words are important. Words are the signs of ideas and react upon ideas. Inadequate words, if they are in current usage, have the privilege of making false ideas invulnerable. Let us avoid using the word "psychology" in the singular and in a non-qualified way. Let us avoid the use of vague or arbitrary modifiers accidentally popularized by the success of a book or a professor (general psychology, dynamic psychology, . . .). To the greatest extent possible, and at least in regard to the principal divisions of education, let us require the use of terms designating precisely defined epistemological essences. Moral psychology, philosophical psychology, applied psychology are expressions whose widespread use would contribute a lot to making the situation less confusing.

2. With regard to the relation between positive psychology and philosophical psychology in the organization of instruction, two tendencies are constantly apparent since positive psychology has become aware of its possibilities: the tendency to maintain the two disciplines united and the tendency to separate them. In spite of the claims for autonomy often uttered by positive psychologists, the first tendency rather generally continues to prevail. In a large number of universities and colleges, the teaching of positive psychology remains more or less closely connected with the teaching of philosophy. It is quite unreasonable to let things

go on at the mercy of prejudices and fashions. It is indispensable to take a stand in favor of either of these two tendencies.

The epistemological principles that we have set out favor the tendency toward separation. But the issue is complicated by a far-reaching historical accident; it is a fact that positive psychology as it is currently taught has an annoying propensity to turn itself into an instrument of different philosophies, acknowledged or not, generally stupid and harmful, but very attractive because of the prestige conferred upon them by their association with positive science. As it is beyond the power of anyone to bring these insidious philosophical influences to an end, conscientious educators are inclined to think that it is convenient to keep the teaching of positive psychology under the control of philosophers, and that a good way to ensure this control is to treat institutes of positive psychology as appendages to departments of philosophy.

It is not certain that this is a good method. The uniting of the teaching of positive psychology with philosophy perpetuates the confusion which constitutes all the strength of philosophies hidden behind the appearances of positive knowledge. Positive psychology and applied psychology will be much less tempted to pass themselves off as philosophies, ethical systems, and forms of wisdom, as the organization of instruction will more clearly reveal their true nature by making them take their place among other positive and applied sciences.

3. Finally, we express the wish that moral psychology be recognized by the programs as a distinct discipline and that it be taught in the departments of philosophy. The systematization of moral psychology by itself would constitute a great scientific advance. This progress would be particularly timely in an era in which the knowledge of moral man is so seriously threatened by the technocratic imperialism of different neo-positivist groups. As an indication of this, let us refer to an encouraging experience. During the academic year 1940–1941, I had the occasion to pursue a series of inquiries on the problem of moral psychology with a group of graduate students at the University of Notre Dame. Here is the list of topics that we adopted. (1) Man and Nature; (2) Man at Work; (3) Property; (4) Play; (5) Authority; (6) Love and Family Life; (7) Man in the Face of Death.

As most of the questions were dealt with in the form of student presentations and discussions in which the professor played a subdued role, I may be allowed to say that these inquiries were pursued with extraordinary interest. By observing the reactions of my young companions, I understood that our studies in moral psychology gave them the rare satisfaction of instilling new life into their philosophical thinking by nourishing them with what was most vital in their human experience.

Notes

1. See Simon, *Freedom of Choice* [ed. note].
2. In his introduction to Georges Dumas's *Traité de psychologie*, 2 vols. (Paris: Alcan, 1923–1924), André Lalande says that the psychologist can expect services from literary documents only analogous to those provided by anatomical color plates in the study of organisms. That is no doubt true when it concerns positive psychology. But in moral psychology the great artist exercises a unique function of discovery and expression admirably described by Stanislas Fumet in the following lines: "The great artist succeeds in showing us what is hidden in the abysses of the human heart since the loss of the Earthly Paradise. These age-old treasures dwell in each of us as they do in the artist, but he has a wonderful pulley to bring them to the surface that others do not possess. Shakespeare, Baudelaire, Dostoevsky, Wagner, or Bloy who discover unexpected stars in the inverted firmament of souls do not have more constellations in them than the ordinary pedestrian crossing the street. But they have endless wands to capture them." *Mission de Léon Bloy* (Paris: Desclée de Brouwer, 1935), p. 12.
3. Cf. Ludwig Klages, *The Science of Character*, trans. W. H. Johnston (London: Allen and Unwin, 1929), especially Chapter 1, entitled "School Psychology and Its Relation to Characterology"; and Gustave Thibon's work on Klages, *La science du caractère* (Paris: Desclée de Brouwer, 1934).
4. The idea that it is impossible to detach positive science from philosophy when the subject under investigation is the human soul has been vigorously expressed by Maine de Biran, whose pioneering works have had a considerable influence on the movement of psychology during the last two generations. "Physicists can limit themselves to observing external facts, phenomena within their province, and grasping their connection or order of succession based on experience. They presuppose the absolute reality of causes and substances. They do not need to determine their nature nor to inquire on what grounds we know or believe these realities. But in psychology, even of the most experimental kind, it is hardly possible to disregard the efficient cause of certain phenomena,

to disregard the fact that this cause, insofar as it is originally identified with the *self*, becomes itself the subject of the science; that its acts and their immediate outcomes constitute an essential part of the internal phenomena—In short, that the notions of forces, of enduring substances, and the invincible belief in their reality, are placed in the first rank of facts and make up the primary elements of social science, from whence it follows that they cannot be disregarded without completely distorting the very subject of the study that was proposed. . . ." *Œuvres complètes* X, ed. Pierre Tisserand (Geneva-Paris: Slatkine, 1982), pp. 76–77.

5. There is nothing conclusive about the texts relied on by certain interpreters of Aristotle who consider that the study of the rational part of the soul in the Aristotelian system belongs to metaphysics, not to the philosophy of nature. Compare Eduard Zeller, *Die Philosophie der Griechen in ihrer geschichtlichen Entwicklung*, 5 vols. (Leipzig: Fues, 1869–1881), II, 2, who asserts that it belongs to metaphysics with St. Thomas and his great commentators who assert, on the contrary, that the definition of the soul as "the initial act of an organized body" includes all the parts of the soul and that consequently the entire treatise *De anima* depends on the philosophy of nature. Cf. St. Thomas, *Comm. in Physica*, II, lect. 4; Cajetan, *Comm. in De anima*, I, text 15; John of Saint Thomas, *Philosophia naturalis*, I, 9, a. s, *Cursus philosophicus* II, pp. 180ff.

6. Many authors systematically confuse positive psychology and applied psychology. But it is easy to notice that in psychology as elsewhere a complete polarization of the science by its applications has the effect of arbitrarily limiting its theoretical fruitfulness and, in the final analysis, its practical fruitfulness.

7. Cf. for instance Franz Alexander, *Our Age of Unreason: A Study of the Irrational Forces in Social Life* (Philadelphia: Lippincott, 1942), p. 22.

8. See Simon's *Definition of Moral Virtue* [ed. note].

Bibliography

Alexander, Franz. *Our Age of Unreason: A Study of the Irrational Forces in Social Life.* Philadelphia: Lippincott, 1942.
Aquinas, Thomas. *Commentary on Aristotle's On Interpretation (Peri Hermeneias).* Trans. Jean T. Oesterle. Milwaukee: Marquette University Press, 1962.
———. *The Division and Methods of the Sciences.* Trans. Armand Maurer. Toronto: Pontifical Institute of Mediaeval Studies, 1986.
———. *On the Truth of the Catholic Faith: Summa contra Gentiles* I. Trans. Anton C. Pegis. Garden City, N.Y.: Doubleday, 1955.
———. *On the Truth of the Catholic Faith: Summa contra Gentiles* III. Trans. Vernon J. Bourke. Garden City, N.Y.: Doubleday, 1956.
———. *Summa totius logicae Aristotelis.* In *Opuscula omnia, genuina quidem necnon spuria melioris notae debito ordine collecta cura et studio.* Ed. Pierre Mandonnet. 5 vols. Paris: Lethielleux, 1927.
———. *Truth* III. Trans. Robert W. Schmidt. Chicago: Regnery, 1954.
Aron, Raymond. *Main Currents in Sociological Thought.* Trans. Richard Howard and Helen Weaver. 2 vols. London: Weidenfeld & Nicolson, 1967.
Bossuet, Jacques-Bénigne. *Discourse on Universal History.* Trans. Elborg Forster. Chicago: The University of Chicago Press, 1976.
Broglie, Louis de. *Matter and Light: The New Physics.* Trans. W. H. Johnston. New York: Dover, 1946.
Bulletin bibliographique de la Société des Écrivains Canadiens. Montreal: Société des Écrivains Canadiens, 1948, 1949.
Cahalan, John C. *Causal Realism: An Essay on Philosophical Method and the Foundations of Knowledge.* Lanham, Md.: University Press of America, 1985.
Carnap, Rudolf. "The Elimination of Metaphysics." In *Logical Positivism.* Ed. A. J. Ayer. Glencoe, Ill.: Free Press, 1980. Pp. 60–81.
———. *Philosophy and Logical Syntax.* London: Kegan Paul, 1935.

Castelnuovo, Guido. *La probabilité dans les différentes branches de la science*. Paris: Hermann, 1937.

Compton, Arthur H. *The Freedom of Man*. New Haven, Conn.: Yale University Press, 1935.

Cournot, Antoine Augustin. *An Essay on the Foundation of Our Knowledge*. New York: Liberal Arts, 1956.

———. *Matérialisme, vitalisme, rationalisme*. Rev ed. Paris: Hachette, 1923.

———. *Traité de l'enchaînement des idées fondamentales dans les sciences et dans l'histoire*. Rev. ed. Paris: Hachette, 1911.

Dalbiez, Roland. *Psychoanalytical Method and the Doctrine of Freud*. 2 vols. London and New York: Longmans, Green, 1941. Repr. Freeport, N.Y.: Books for Libraries, 1972.

De Koninck, Charles. "Réflexions sur le problème de l'indéterminisme." *Revue Thomiste*, 43 (1937), 227–52, 393–409.

Dent, N. J. H. *The Moral Psychology of the Virtues*. Cambridge: Cambridge University Press, 1984.

Dingle, Herbert. *Science and Human Experience*. New York: Macmillan, 1932.

Duhem, Pierre. *Le système du monde: Histoire des doctrines cosmologiques de Platon à Copernic*. 10 vols. Paris: Hermann, 1913.

Eddington, Sir Arthur S. *The Nature of the Physical World*. London: Dent, 1947.

L'Édition littéraire au Québec de 1940 à 1960. Ed. Richard Giguère and Jacques Michon. Sherbrook, Quebec: Université de Sherbrooke, 1985.

Einstein, Albert. "Physics and Reality." *Ideas and Opinions*. New York: Crown, 1954. Pp. 290–323.

Fellows, Otis E., and Milliken, Stephen F. *Buffon*. New York: Twayne, 1972.

Frank, Phillip. "Berichte: Die Prager Vorkonferenz 1934." *Erkenntnis*, 5 (1935), 5.

———. *Between Physics and Philosophy*. Cambridge: Harvard University Press, 1941.

———. *La fin de la physique mécaniste*. Paris: Hermann, 1936.

Freud, Sigmund. *The Psychopathology of Everyday Life*. London: Hogarth, 1960.

Fumet, Stanislas. *Mission de Léon Bloy*. Paris: Desclée de Brouwer, 1935.

Gammon, Francis L., Jr. "The Philosophical Thought of Yves R.

Simon: A Brief Survey." *Revue de l'Université d'Ottawa,* 42, No. 2 (1972), 237–44.

Garrigou-Lagrange, Réginald. *Le sens du mystère et le clair-obscur intellectuel: Nature et surnaturel.* Paris: Desclée de Brouwer, 1934.

Goblot, Edmond. *Essai sur la classification des sciences.* Paris: Alcan, 1898.

Gurwitsch, Aron. Review of Yves R. Simon, *Prévoir et savoir. Philosophy and Phenomenological Research,* 7, No. 3 (March 1947), 339–42.

Hahn, Hans. "Logic, Mathematics and Knowledge of Nature." In *Logical Positivism.* Ed. A. J. Ayer. Glencoe, Ill.: Free Press, 1959. Pp. 147–61.

———. *Logik, Mathematik und Naturerkennen.* Vienna: Gerold, 1933.

———. *Logique, mathématique et connaissance de la réalité.* Paris: Hermann, 1934.

John of St. Thomas. *Cursus Philosophicus: Philosophia Naturalis.* Ed. Beatus Reiser. Turin: Marietti, 1933.

Juvet, Gustave. *La structure des nouvelles théories physiques.* Paris: Alcan, 1933.

Klages, Ludwig. *The Science of Character.* Trans. W. H. Johnston. London: Allen & Unwin, 1929.

Kuhn, Thomas. *The Structure of Scientific Revolutions.* Chicago: The University of Chicago Press, 1962.

Lalande, André. Introduction to Georges Dumas, *Traité de psychologie.* 2 vols. Paris: Alcan, 1923–1924.

Laplace, Pierre-Simon. *A Philosophical Essay on Probabilities.* New York: Dover, 1951.

Maine de Biran, Pierre. *Œuvres complètes.* Ed. Pierre Tisserand. 9 vols. Geneva-Paris: Slatkine, 1982.

Marcel, Gabriel. "On the Ontological Mystery." *The Philosophy of Existentialism.* New York: Citadel, 1963. Pp. 9–46.

Margolin, Jean-Claude. *Bachelard.* Paris: Seuil, 1974.

Maritain, Jacques. *Distinguer pour unir, ou, Les degrés du savoir.* Paris: Desclée de Brouwer, 1932.

———. *Distinguish to Unite, or, The Degrees of Knowledge.* Trans. Gerald B. Phelan. New York: Scribner's, 1959.

———. *Philosophy of Nature.* New York: Philosophical Library, 1951.

———. *A Preface to Metaphysics.* New York: Sheed & Ward, 1939. Repr. Freeport, N.Y.: Books for Libraries, 1979.

———. "Reflections on Necessity and Contingency." In *Essays on Thomism*. Ed. Robert E. Brennan. New York: Sheed & Ward, 1942. Repr. Freeport, N.Y.: Books for Libraries, 1972. Pp. 27–37.

———. "Sign and Symbol." *Ransoming the Time*. New York: Gordian, 1972. Pp. 218–26.

The Material Logic of John of St. Thomas. Trans. Yves R. Simon, John J. Glanville, and G. Donald Hollenhorst. Chicago: The University of Chicago Press, 1955. Repr. 1965.

The Meditations of Marcus Aurelius Antoninus. Trans. A. S. L. Farquharson. Oxford: Oxford University Press, 1990.

Meyerson, Émile. *De l'explication dans les sciences*. 2nd ed. Paris: Payot, 1927.

———. *Identity and Reality*. Trans. Kate Loewenberg. New York: Dover, 1962.

Milhaud, Gaston. "Le hasard chez Aristote et chez Cournot." *Revue de Métaphysique et de Morale*, 10 (1902), 667–81.

Mill, John Stuart. *A System of Logic: Ratiocinative and Inductive*. Toronto: University of Toronto Press, 1973.

Orenstein, Alex. *Willard Van Orman Quine*. Boston: Twayne, 1977.

Perrault, Marcel. Review of Yves R. Simon, *Prévoir et savoir*. *Revue de l'Université d'Ottawa*, 15, No. 1 (1945), 248–49.

Piaget, Jean. *Sagesse et illusions de la philosophie*. Paris: Presses Universitaires de France, 1965.

Piéron, Henri. "Essai sur le hasard: La psychologie d'un concept." *Revue de Métaphysique et de Morale*, 10 (1902), 682–95.

Planck, Max. "Causality in Nature." In *The Philosophy of Physics*. Trans. W. H. Johnston. New York: Norton, 1936. Pp. 43–83.

———. "Causation and Free Will." *Where Is Science Going?* Trans. James Murphy. New York: AMS Press, 1977. Pp. 107–40.

Poincaré, Henri. *Science and Method*. Trans. Francis Maitland. New York: Dover, 1952.

———. *The Value of Science*. New York: Dover, 1958.

———. *La valeur de la science*. Paris: Flammarion, 1904.

Poirier, René. "Autour de Bachelard épistémologue." *Colloque de Cerisy-La-Salle sur le thème Gaston Bachelard, 1970*. Paris: Union Générale d'Éditions, 1974.

Rabeau, Gaston. *Dieu: Son existence et sa providence*. Paris: Bloud et Gay, 1933.

Schlick, Moritz. "Positivism and Realism." In *Essential Readings in Logical Positivism*. Ed. Oswald Hanfling. Oxford: Blackwell, 1981. Pp. 83–110.

Schrödinger, Erwin. *Science and the Human Temperament*. Trans. James Murphy and W. H. Johnston. New York: Norton, 1935.

Simon, Yves R. "Causality and Indetermination." Lecture 16, Committee on Social Thought, University of Chicago, February 26, 1969.

———. "Chance and Determinism in Philosophy and Science." In *The Great Dialogue of Nature and Space*. Ed. Gerard J. Dalcourt. Albany, N.Y.: Magi, 1970. Pp. 181–204.

———. *The Definition of Moral Virtue*. Ed. Vukan Kuic. New York: Fordham University Press, 1986. Repr. 1989.

———. "An Essay on Sensation." In *Philosophy of Knowledge: Selected Readings*. Ed. Roland Houde and Joseph P. Mullally. Chicago: Lippincott, 1960. Pp. 55–95.

———. *Freedom of Choice*. Ed. Peter Wolff. New York: Fordham University Press, 1969. Repr. 1987, 1992.

———. "From the Science of Nature to the Science of Society." *The New Scholasticism*, 27, No. 3 (July 1953), 280–304. Repr. in *Practical Knowledge*. Ed. Robert J. Mulvaney. New York: Fordham University Press, 1991. Pp. 115–36.

———. *The Great Dialogue of Nature and Space*. Ed. Gerard J. Dalcourt. Albany, N.Y.: Magi, 1970.

———. "How We Explain Nature." In *The Great Dialogue of Nature and Space*. Ed. Gerard J. Dalcourt. Albany, N.Y.: Magi, 1970. Pp. 21–35.

———. *An Introduction to Metaphysics of Knowledge*. Trans. Vukan Kuic and Richard J. Thompson. New York: Fordham University Press, 1990.

———. *La marche à la délivrance*. New York: Éditions de la Maison Française, 1942.

———. *The March to Liberation*. Trans. Victor M. Hamm. Milwaukee: Tower, 1942.

———. "Maritain's Philosophy of the Sciences." In Jacques Maritain. *Philosophy of Nature*. New York: Philosophical Library, 1951. Pp. 152–82. Repr. in *The Philosophy of Physics*. Ed. Vincent E. Smith. Jamaica, N.Y.: St. John's University Press, 1961. Pp. 25–39.

——. "On Order in Analogical Sets." *The New Scholasticism*, 34, No. 1 (January 1960), 1–42.

——. "On the Foreseeability of Free Acts." *The New Scholasticism*, 22, No. 2 (October 1948), 357–70.

——. "Philosophy of Science." *Revue de Philosophie*, N.S. 6, No. 1 (January 1935), 53–64.

——. *Practical Knowledge.* Ed. Robert J. Mulvaney. New York: Fordham University Press, 1991.

——. Review of Gaston Bachelard, *L'expérience de l'espace dans la physique contemporaine. Revue de Philosophie*, N.S. 8, No. 4 (July-August 1937), 355.

——. *The Tradition of Natural Law: A Philosopher's Reflections.* Ed. Vukan Kuic. New York: Fordham University Press, 1965. Repr. 1967. Rev. ed. 1992.

Sorel, Georges. *D'Aristote à Marx.* Paris: Rivière, 1935.

——. *Les préoccupations métaphysiques des physiciens modernes.* Paris: Cahiers de la Quinazine, 1907.

——. *Reflections on Violence.* Trans. T. E. Hulme. New York: Collier, 1961.

Thibon, Gustave. *La science du caractère.* Paris: Desclée de Brouwer, 1934.

Toulmin, Stephen. *Foresight and Understanding: An Inquiry into the Aims of Science.* New York: Harper Torchbook, 1963.

——. *Return to Cosmology: Postmodern Science and the Theology of Nature.* Berkeley: University of California Press, 1982.

Vouillemin, Général Charles Ernest. "Hasard et régularité." *Revue d'Artillerie* (1937).

——. "Le positivisme critique de l'École de Vienne." *Revue de Philosophie*, N.S. 6, No. 3 (1935), 229–54.

——. "Sur la theorie physique." *Revue d'Artillerie* (March 1932).

Whitehead, Alfred North. *Science and the Modern World.* New York: Macmillan, 1928.

Zeller, Eduard. *Die Philosophie der Griechen in ihrer geschichtlichen Entwicklung.* 5 vols. Leipzig: Fues, 1869–1881.

Index

Absolute beginning, 17
Abstraction:
   abstractio formalis;—totalis; 103, 108
   contingent situations, 108
   differentiation process, 93
   distinction of degrees, 97–98n5
   experimental material, 69
   extensive, 103
   first order of, 100
   moral psychology, 116
   physical definitions, 105
   physical knowledge, 103
   three orders of, 97n5
   two terms: what; from which, 102
   typological, 102, 103
   universals, 108
Accidents:
   predictable, 7
   responsibility for, 10
Acosmism, 13, 16
Action:
   activity (concept), 24
   emanation; producing effect, 19
   formal predetermination, 19, 36
   proceeding from an agent, 19, 55n22
Actuality, 17, 18; *see also* Being; Reality
Adler, Mortimer J., xviii$n$11
Adoration, 15
Agent:
   principle(s) of causality, 19
   produces something similar to itself, 55n22
Analogy of being, 92, 95
Anatomy, 22
Anscombe, G. E. M., xvii
Anxiety neurosis, 120
Aquinas, *see* Thomas Aquinas, Saint
Architecture, 71
Aristotelian school, 4
Aristotle (Aristotelianism), 4, 13, 16, 29, 50n6, 71n4, 90, 92, 94, 98n5, 100, 105, 106, 118, 129n5
   on material causality, 17
   on potency, 53n19
   physics, 59n39
   principle of causality, 12
   treatise on the soul, 112
   words as sign, 75–76
Arithmetic, 98n5
Arm, anatomy of, 22
Art:
   analogies with philosophy, 91
   matter and form, 55n22
   nature as dramatic art, 65n62
   order of direction, 100
Assent, universal, 1, 2
Astronomical events, 63n54
Astronomy: popular nightmares, 63n54
Atom, 30, 31, 57n32
Augustine, Saint, 112

Bachelard, Gaston, xii, xix$n$20
Balzac, Honoré de, 112
Baudelaire, Charles, 128n2
Becoming:
   ability to change shape, 17
   novelty, 17, 20
   plurality in nature, 16
Beginning: absolute, 17
Behavior:
   control over self, 124
   man of action, 114
Being(s):
   absolute rejection of, 48
   actual being; being in potency, 18
   analogical character of, 95
   bond connecting thought to, 95
   capacity to cease being, 104
   corruptible, 104
   created existences are contingent, 25–26
   equivalence: *ens mobile* and *ens materiale,* 105
   essential unity, 12
   incorruptible formulation, 104

138   INDEX

intelligibility, 94
intuition of, 95
language in philosophy, xi
meaning reduced to, 76
measurement in science and, 87
mobile, sensible, material, 98n5, 100
mode imposed by nature, 108
necessity and contingency in, 39
nonontological systems, 39
object of intellect, 101
ontological interpretation of facts, 119n3
ontological types, 109
possibility to act, 18
primary law: self-identity, 101
problem of origins, 17, 18
proper causes, 53n16
rational appetite for, 119n2
science in relation to, 47
unity and multiplicity in, 18
unity in diversity, 95
Bergson, Henri, xvi, 91
Biologists: on evolutionary processes, 66n62
Bloy, Léon, 128n2
Bossuet, Jacques-Bénigne, 53n16
Boyle's law, 28, 58n38, 70
Broglie, Louis de, 41
Buffon, Georges-Louis Leclerc de, 20
Business psychology, 113

Cahalan, John C., xviii$n$8, xxi
Cajetan, 99n5
Carnap, Rudolf, xiv, 77, 78, 95
Cassirer, Ernst, xii
Castelnuovo, Guido, 34
Catastrophes, 66n62
Categories: Aristotelian, 109
Cause (-s- ality), ix, 1, 2, 39
  accidental, 20
  accidental existence, 12
  aggregate of; convergence of, 11
  Aristotelian/Thomistic theory, 16
  case: explanation of an accident, 14–15
  chance, see Chance
  characterology, 117
  concept independent of sense-perception, 46
  conceptual transformations, 21
  conservative position, 32

determined:
  free: two systems in the soul, 122
  psychological techniques, 123
  reactions, 122
divine concursus; secondary causes, 25
divine providence and, 52n16
duality of cause and caused, 18
every event explained by, 12
explanation of the universe, 26
explanatory knowledge, 101
free causality, 115
freedom from causal dependence, 72n5
future contingents, 61n54
identity and, 16–20, 55n22
interdependent/independent series in, 9
irreducible plurality of, 12, 13
knowledge through and in, 14
material, 17, 23
motion principle, 19
necessary law of being, 32
nonunified proper causes; rolling dice, 35–36
ontological concept, 21
otherness of the cause, 18
physical events, 67–68
Planck on, 45
plurality of, unified, 15–16
predetermination of action and effect, 19, 20, 36
principle of, xi
principle of finality, 19
principle of proper causality, 19
proper causality in non-ontological knowledge, 21–24
proper cause (concept), 20
real unity and, 12
regularity; joint operation 1, 10
relation between two magnitudes, 24
scientific causality, 55n22
specialized branches of science and, 45
tendencies and, 58n35
theological causality, 55n22
unity; work of reason, 12–13
universal validity of the idea, 39
unpredictability of chance, 7
see also Effect; Explanation
Cause and effect:

INDEX

essential tendency of a thing, 26
providence and chance occurrences, 52n16
symmetry, 34
Certainty, 1, 44, 46, 60n48, 63n54, 77, 79, 101, 111
Chance (chance events), 2, 7–10, 39
  absolute, 11
  Cajetan on, 53n16
  causal identification, 13
  common sense interpretation, 8, 10
  common/practical notion of, 50n7
  concurrence of two or more causes, 52n16
  Cournot on, 50nn6&7
  definition, 9
  denial by classical physicists, 31
  displaced by intention, 53n16
  essential prediction, 63n54
  event without a real cause, 12
  events occur inevitably, necessarily, determinately, 11
  example; analysis, 14–15
  future contingents, 62n54
  growth; discrimination of plurality, 28
  integration in the law, 29
  law against events of, 30
  meaning for philosopher and scientist, 87
  nonunified plurality of causes, 10
  not of itself unforeseeable, 68
  predetermination of, 12
  predictability, 9, 44, 62n54
  primitive; subjective, 32
  promotion of, 28
  providence and, 53n16
  reality of, 15–16
  root of causality, 31
  unexplainable, unintelligible, nonrational, 13
  unity; work of reason, 12–13
  universal intelligibility, 13
Change:
  philosophical interpretation, 39
  physical, xi
  potency, 17
  reality of, 13
  substantial, 98n5
  universal, 29
Character, 114
Characterology, 117

Chemical elements: causes of the compound, 23
Chemical reaction, 23
Christian life, 94
Clericalism, political, 94
Cognition:
  object "unimaginable by nature," 84
  two words in different languages, 81
Coincidence, 53n16
Common sense, xviiin11, 7
  determinism, 40
  identification; sensation, 105
  illusions of, 15
  interpretation of chance, 10
  meaning, 8
  origin of science and philosophy, 8
  psychological facts, 118–19
  regulating concept, 39
Communicability, 1
Compton, Arthur H., 42
Compton effect, 43n4
Comte, Auguste, xii, xiii, 51n15, 67, 69, 72–73n12, 119–20
Concepts:
  definition of word (sign), 76
  intentional sign, 75
  objective; subject and predicate, 82
  resolution, xi
Conditioning: causality and, xi
Condorcet, Marquis de, 66n62
Conduct:
  foreseeing, 114, 116
  free will, 115
Constraint, psychic, 124
Contemplation, 71n1
Contingency (the Contingent):
  being can be other than it is, 87
  created existence, 64n54
  denial by classical physicists, 31
  explanation, 108
  grounded in necessity, 11
  individually existing things, 100
  influence of First Cause, 15
  miraculous intervention, 63n54
  natural events, 39
  necessity of, 11
  prediction, 61–62n54
  will of God, 52n16
  see also Being; Future contingents; Necessity

Contradiction, law of, 80
Correlative things, 53n16
Corruption, 104
Cosmic image, 49, 65, 66n62
Cosmology:
 Aristotle's treatises on, 98n5
 Thomism, 93
Cournot, Antoine Augustin, 9, 14, 50nn6&7
Cratylus, 29
Creatures:
 divine government, 15
 mobile being, 104
Critical realism, xii, 84
Critique of knowledge (term), xii

Dalbiez, Roland, 57n35
Darwin, Charles, 49
Deduction:
 Hahn on, 81
 mathematical, 82
 tautology, 81
Definition:
 literary philosophy, 4
 philosophy and science, x
 physical, 105
 sensible beings, 109
Degrees of Knowledge, The (Maritain), xii, 3, 84
De Koninck, Charles, 60n48, 110n4
Demonstration: conclusions; concern of, 57n33
Dent, N. J .H., xvii
Derrida, Jacques, x
Desaix de Veygoux, Louis Charles Antoine, 9
Descartes, René, 118
Description, 109
Desire: technology and, 125
Despotism, 125
Determinability: matter and form, 110n5
Determination:
 empiriometrical, 47
 intrinsic lack of, 104
 man of action, 113
Determinism(s), ix, x, 1, 2, 7–66
 absolute, 40
 certain and exact prediction, 40, 46
 certainty pertaining to a whole, 60n48

diverse philosophical standpoints, 38
essential necessity, 11
exact calculation, 87
Frank and, 87
idea in many conceptual systems, 40
immediate foundation of concept of, 19
lawful processes, 68–69
macroscopic, 60n48
mathematico-physical meaning of, 87
metaphysical character of, 31, 46
microscopic indeterminism, 60n48
natural, 115
ontological, 46
philosophic ambiguity, 10
positive notion of, 40
question for philosophic criticism, 40
three systems of, xi
see also Free . . . ; Indeterminacy
Die (dice), 34–38, 60n48
Differentiation, process of, 93
Dimensions, 86
Dingle, Herbert, 43
Discovery, 30
Disorder: phase of, 38
Dissent: technology and, 125
Distinction:
 Aristotelian philosophy, 4
 inconstancy and, 109
 objective aspects of a thing, 26
 subject and predicate, 82
 virtual distinction, 16
Diversity:
 efficient cause, 18
 homogeneity, 52n15
 nonbeing in things, 16
 rationalism, 13
 safeguarding unity, 19
 unity and, 54n21
Divine concursus: secondary causes and, 25
Divine government, *see* Providence
Division, 93, 95, 102, 105, 126
Dostoevsky, Fedor, 112, 128n2
Dread, 16
Duhem, Pierre, xii, 59n39

Eclipse, 63n54

Eddington, Sir Arthur, xiii, 41
Education:
  division of sciences and, 126
  principal divisions of, 126
Effect:
  chance occurrence in regard to cause, 52n16
  explanation by cause, 18
  likeness of, in proper cause, 20
  predetermination in the cause, 19, 36
  predetermination of, 20, 36
  see also Cause
Efficiency: relation of, 24
Efficient cause, 13, 17, 23, 24
  complete expression of, 54n21
  explanation; otherness, 18
  form of the effect, 54n22
  Maine de Biran, Pierre, 128n4
Einstein, Albert, 83, 86
Elasticity, 68
Eleatics, 13
Electron:
  freedom, 88
  momentum, 42–43
  position in space, 43
Eminence, 99n5
Emotional makeup, 78
Empirical science, 77
Empiricism:
  Hahn on, 79
  new, xiv
  tautology, 82
Empiriological thought, 94, 95
Empiriometry, 47
End(s):
  attaining, 14
  principle of finality, 19
Engineering, xv, 11
Ens materiale, 109n1
Epistemology, 1, 21, 46–47, 125–26
  Bergsonian, 91
  Cartesianism, 92
  deterministic, 117
  meaning: theory of science, 90
  pluralism in, 90–99
  scientistic, 92
  univocity, 91, 92
Equality: newborn child and monkey, 54n19
Equations, 24
Equilibrium, 68

Essence: scientific object, 67
Eternity, 26
Ether, elasticity of, 73n12
Ethics:
  neutrality; social science, xv
  normative, 78
  psychological concepts relevant to, xvii
  theoretical meaning, 79
  validity of moral knowledge, 91
  Vienna Circle, 78
Events:
  causal series, 67–68
  foreseen in its cause, 43–44
  historical event: contingency, 101
  predictability of physical events, 7
  see also Fortuitous events
Evidence:
  experience and reason, 1
  possessed by intelligible beings, 77
Evil, see Good and evil
Evolution: origin of modern ideas, 66n62
Excluded middle, law of, 80
Existence:
  beginnings, 17
  created; radical contingency, 64n54
  one necessary: divine existence, 25
  outside the mind, 27
  see also Being
Existential states: scientific knowledge, 25
Experience:
  daily; practical, 106
  evidence; understanding, 1
  immediate data of, 102
  propositions of logic and mathematics, 80
  scientific, 27
  theoretical psychology, 117
Experimental absolute, xii–xiii, xixn22
Experimental psychology (the term), 121
Experimental science: defined, 72n10
Explanation, ix, 19, 108
  ambition of science, 11
  causation and, xi
  common search for, x
  efficient cause of events, 13
  identification, 16, 51n15
  intelligibility of the world, 16

law alone, without cause, 51n15
material causality; change, 17
perfect knowledge, 114
prediction and, xiii
proper causality, 21
science and philosophy, xi
sufficient number of causes needed, 12
theoretical; practical, 14, 15
universe reduced to single cause, 26
wholes presented by experience, 117
Extinct species, 103–104

Factory: industrial work output, 124
Facts:
  ontological interpretation, 119n3
  predictability, 10
  scientific, xixn22
  systematization, 119n4
Faith, 95
Falsification, ix
Feyerabend, Paul, viii
Finality:
  predetermination to acting, 19
  theory of universal motion, 20
First Cause:
  infallibly effective, 15–16
  preternatural intervention by, 72n9
Forecast, see Prediction; Science
Foreseeing (-knowing- sight), 61–62n54
  essential function of science, 71n1
  free events, 62n54
  natural events, 43–44
  scientific knowledge, 67
Form:
  irreducible obscurity, 110n5
  principle of stability, 108
  quality proceeds from, 108
  sensible qualities, 108, 109
  see also Matter
Formulas: statistical or causal, 31
Formulation: types; individuation, 105
Fortuitous events:
  aggregate of causes, 11
  Cajetan on, 53n16
  common sense psychology, 15
  influence of First Cause, 15
  plurality of individuals, 27
  providence and, 52n16
  results of chance, 9, 10
  two things in motion, 11
  see also Chance
Fowlie, Wallace, vii
France, vii, viii
Frank, Philipp, xiii, 64n60, 86, 91
Free choice, xvii, 114–15, 124
  denial of, 122
  theoretical psychology, 117
Free will:
  conduct and, 115
  Frank and, 87
  goodness of, 27
  influences on, 124
  propaganda and, 124
  understanding man's free acts, 122
Freedom:
  denial by classical physicists, 31
  modifying dispositions of, 122
  proper causality, 55n22
  science and philosophy, 88
  suppression of, 122, 124, 125
Fresnel, Augustin-Jean, 73n12
Freud, Sigmund, xvi, 22, 36, 56n28
Fumet, Stanislaus, 128n2
Future and past, 20
Future contingents:
  foreseeing, 61–62n54
  foreseen with certainty by accident, 63n54
  Maritain on, 61–62n54
  scientific knowledge of, 62n54
Future perfect (velocity), 42

Garrigou-Lagrange, Réginald, 94
Gaseous mass, 28–29
Genus, 103
Geological age, 66n62
Geometrical progression, law of, 32
Geometry, 59n39, 70, 98n5
Gnoseology, 90, 91
Goblot, Edmond, 71n4
God:
  author of nature, 16
  bothering, 63n54
  cause of Himself, 18
  existence identical with essence, 25
  immutable, 104
  see also First Cause
Good and evil, 115, 116
Goodness, 27
Gravity, 83

Gurwitsch, Aron, xii

Hahn, Hans, xiv, 79ff.
Hamlet, 116
Hand, anatomy of, 221
Healing, praying for, 63n54
Heart: hidden in abysses of, 128n2
Hegel, G. W. F., 94, 96n3
Heisenberg, Werner, xiii
Heraclitus, 29
Hiring drivers: case; applied psychology, 123n2
Homogeneity: diversity, 52n15
Honesty, 114
Honor, 113, 114
Human development, 91
Humans: empiriological definition, xi
Hume, David, 31, 32
Hypnotism, 124–25
Hypotheses:
   Comte on, 72–73n12
   experimental method, 72n10
   explanations in science, 57n32

Idealism, xii, 3, 65–66n62, 78
Ideas:
   Augustinian theory of, 71n4
   co-ordination of, 97n4
   metaphysical character, 94
   realization of, 19
   studied in relation to reality, 39–40
   word as sign of, 126n1
Identity(-ification):
   causal; perfection, 26
   causality and, 16–20
   cause reduced to relation of, 55n22
   chemical studies, 23
   connection and continuity, 18
   explanation and, 16
   Parmenides on, 13
   proper causality, 19
   Quine on, x
   scientific explanation, 51n15
   search for unity, 17
   substantial form, 107
Ignorance, 92
Imagination:
   Freud, 56n28
   positive science, 48–49
   truth, 65n62
Impermanence, 109
Inanimate individual, 27

Inconstancy, 109
Incorruptible things, 57n33
Indeterminacy (-ism):
   cause, 43n2
   critical judgment, 40
   Frank and, 87
   irremovable, 41
   man/nature relationship, 44
   microphysical observation, 43n1
   philosophic criticism; modern physicist, 40
   philosophic meaning, 47
   prediction of position of particle, 41–42
Individuality:
   law of, 27
   living, inanimate things, 27
   practical notion of, 27
Individuals:
   plurality of, 27
   universal type in, does not change, 29
Individuation, principle of, 105
Industrial work output, 124
Information, 117
Instability, 105
Instincts, 91
Intellect:
   foreseeing use made of, 115
   operation of, 96n4
   practical, 100
Intelligence:
   vast (universal comprehension), 49n1
Intelligibility:
   nonontological apprehension of things, 94
   scientific certainties, 25
   universal, 13
Interdependent/Independent series, 9
Interpretation, x, 1, 2, 8, 10, 12, 28, 39, 48, 119n3
Intuition:
   meaning; Carnap, 95
   sensory; certainty, 77
Invisibility, 84

Jesus Christ:
   human intellect, 92, 93
   infused science, 97n4
   knowledge of Him, 92

John of St Thomas, 57n33, 92, 97, 99n5
Juvet, Gustave, 42

Kant, Immanuel, 118
Kléber, Jean-Baptiste, 9
Knowledge:
  abstracting from free choice, 114–15
  all certain knowledge claimed by science, 91
  business psychology, 113
  by inclination, xvii
  certain, 100, 101
  comparative and formal, 70
  critique; determinism, 39
  defective realization, 25
  explanatory; causes or reasons of things, 101
  factual; scientific propositions, 25
  forms of prediction, xiii
  free of error, 25
  future contingents, 62n54
  imperfect means of knowing, 110n4
  logical positivism, xiv
  metaphysical study of, 121
  monistic conception of, 90
  non-ontological; proper causality, 21–24
  object of scientific, 102, 108
  perfect: certain; explanatory, 92, 101, 114
  physical, *see* Physical knowledge
  physical reality, x
  practical, xv
  qualitatively perfect, 100–101
  realist conception of, 92
  regulating concepts, 39
  Sorel's contribution, 70
  synthetic, 114
  systematic, 67–73
  theoretical thought, 14
  things-in-themselves, 45
  totalistic, 114–16
  validity of, 90–91
  see also Epistemology
Kuhn, Thomas S., viii, 66n63

Laboratory materials, 69
Lackland, John, 85
Lakatos, Imre, viii
Lalande, André, 128n2
Lamarck, Jean-Baptiste, 49
Lamarckism, 66n62
Language:
  criticism, 74, 76
  expressive function; metaphysics, 78, 79
Laplace, Pierre-Simon, 49n1
Law(s):
  essential; statistical, xi, 28
  exact; statistical, 31, 32–33
  Meyerson on, 72n8
  of nature, 80
  physical, 28
  scientific abstractions, 67
Learned persons: superiority of, 92
Leibniz, Gottfried, 118
Leverrier, Urbain-Jean-Joseph, 80, 83
Life insurance, 32
Light:
  corpuscular nature, 43n4
  in chemical reaction, 23
  photons, 42
Linnaean science, 49
Literary criticism: philosophers as critics, x
Literary philosophy, ix, x
Local motion (treatise), 98n5
Locke, John, 79
Locomotion, 99n5
Logic, 97n4, 98n5
  Carnap on logical analysis, 79
  fundamental principles in science, 45
  insufficiencies, 65n62
  logical empiricism, viii
  logical positivism, viii, xiv
  object (Hahn), 80
  statements as tautological, 81, 82
  see also Propositions
Lottery, 35
Love: fortuitous event; case, 15
Lucretius, 118

Mach, Ernst, 83
Maine de Biran, Pierre, xvi, 128n4
Man: measure of all things (positivism), 45
Marengo, battle at, 9
Margolin, Jean-Claude, xix$n$20
Mariotte's law, 58n38
Maritain, Jacques, vii, xii, 3, 84, 93

future contingents, 61–62n54
philosophy within science in, xviiin11
see also DEGREES OF KNOWLEDGE, THE
Marxism, 66n62
Mass points (motion; force), 82
Material cause, 17, 23
Materialism, 3, 17, 72n4, 125
Mathematics:
  object, 75
  Plato, 59n39
  propositions; experience, 80, 82
  relationship to being, 94
  statements as tautological, 81, 82
Matter:
  determinability, 110n5
  inner principle of destruction, 104
  principle of indetermination, 108
  quantity and, 107
Maxwell, James, 80
Meaning:
  daily environment, 106
  empiriological terms, 94, 95
  essential elements of theory of, 75
  meaningless (abused word), 39
  metaphysical propositions, 78
  new empiricism, xiv
  observation related to, 95
  positive scientist; observable data, 76
  predictability and, 10
  relation between sign and object, 76
  science and philosophy, 4
  theory, 74
  verifying, 76
Measurement:
  instruments, 41
  object of, 48
  observable dimensions, 86
  perfecting a method of, 41
  positivism, 45
  science and philosophy, 87
  thing independent of mind, 46n2
Mechanics: classical, 86
Medicine, ix, 98n5
Memory:
  evaluating, 115
  sign of object, 75
Mental health, 125
Mercury, expansion of, 68

Merleau-Ponty, Maurice, xvi
Metaphysics, xi, xiv, 90–91, 95, 97n4, 98n5, 121, 129n5
  archetype of scientific thinking, ix, 91
  Carnap on metaphysicians, 79
  criticism of language, 76
  excluded by empiricist, 77
  expressive function of language, 78, 79
  philosophy of nature, xiv
  specialized branches of science and, 45
Meteorological facts, 63n54
Meyerson, Émile, xii, 3, 13, 16, 17, 49, 51n15, 54n19, 55n22, 72n8
Microphysics, 43
Middle term: truth inferred from, 96n4
Mill, John Stuart, 58n35, 118
Mind:
  ultimate measure of, 46n2
  see also Intellect; Sign
Miracle, praying for, 63–64n54
Mobility, 102, 105, 109, 121
Modernity, 4
Molecule, 29, 31, 83
Monism, xiv, 13, 16
  epistemological, 90
  scientistic, 91
Monkey, newborn, 54n19
Moral feelings, 120
Moral psychology, xvii, 115, 116, 122–23, 124, 126n1
  characterology, 117
  distinct discipline; department of philosophy, 127n3
  great writers and, 128n2
  responsibility (case), 123n2
  scientific systematization, 116–17, 127n3
  value in philosophical thinking, 128
Moralists, 116
Morality:
  applied psychology, 125
  science of the soul important for, xvi
  social science and, xv
Morgan, Thomas Hunt, 66n62
Motion, 18, 29, 99n5
  celestial spheres, 65n62
  described in unified way, 82

determined, 87
finality, 20
hand that rolls the die, 35
principle of causality, 19
result; nonunified forces, 36
Movement:
   concrete, 36
   directive intention, 36
   formal predetermination, 36
   motor scheme, 36, 38
   voluntary; will, 55n22
Multiplication, 81
Music, 95
Musicians, 79
Mutability, 107, 108
Mystery:
   feeling for, 92
   philosophers' involvement with, 91–92
Mystical experience, 92

Napoleon Bonaparte, 101
Nature:
   as idea; as tendency, 20
   concrete reality, 67
   diverse sciences study same nature, 96–97nn
   ends of research about, 67
   Greek view of (dramatic), 65n62
   impossibility of science of, 29
   knowledge/laws of, 80
   law behind natural events, 30
   nonphilosophical science of, 93
   ontological and empiriological knowlecge of, xiv
   ontological interpretation, 48
   philosophy of; science of, 93
   Plato's myths, 59n39
   revealed by regularities, 30
   scientific principle of causality, 44
   statistics; lass of nature, 32
   thought in harmony with, 80
   two laws of, 50n6
   understanding, 26
   unique immanent law, 50n1
Necessity, 39, 40, 87
   abstract possibilities, 67
   contingent realities, 101
   essential, 12
   eternally necessary possiblities, 102, 103, 108
   historical events, 11, 101
   intelligible object, 101
   natural events, 43
   physical, 43
   universal essences, 101
   will of God, 52n16
   see also Contingency
Nerve (anatomy), 22
Nervous disorders, 125
Newton, Isaac, 30–31, 47, 83
Nominalism, 26
Nothing, 17
La Nouvelle Relève, vii
Novelists, 116
Novelty: appearances of, 51n15

Object(s):
   categories, 102
   data of experience in conflict with, 102
   division of sensible objects, 105–106
   essentially divided, 102
   heterogeneity of parts, 107
   homogeneous parts, 107
   incorruptible type, 104
   intelligible; necessary, 101
   knowable; definition, 96n4
   knowledge of, free from error, 25
   law of contradiction, 80
   mutability and materiality, 105
   perception, 106
   philosophy of determinism, 39
   physical; definition, 104
   Plato on, 29
   possible reality, 67
   progress in knowledge of, 82
   qualitative distance in approaches to, xixn22
   quantitative whole; parts, 107
   relations between essences and properties, 67
   scientific:
      essence of, 105
      property of, 108–109
      sensible, 106, 109
   simpliciter (adverb), 104
   simply intelligible, 102
   universal and necessary type, 103
   see also Sign
Observation:
   disturbance affecting prediction, 44
   expressing scientific objects, 48

hypothesis confirmed by, 83
measurement, 95
microphysical, 43
in psychology, 120, 121
thought and, 79, 82
unimaginable by default (Maritain), 84
Omniscience, lack of, 81, 82
Ontological integralism, 94, 95
Ontology: positive psychologist's fight against, 119n2
Operations: pre-existent natures, 19
Opinion: dissent, 125
Order:
  apparent; intelligible, 30–31
  disorder and, 37–38
  root of, 31
  universe; will of God, 52n16
Organic reflex, 20
Otherness, 18

Parmenidean myth, 31
Parmenides, 13, 16
Pascal, Blaise, 112, 116
Past and future, 20
Perception:
  definition, 106
  physical knowledge, xv
Perfection:
  causal, 26
  certainty, 92, 101
  knowledge, 114
  theoretical, 100
Phenomenology, xii
Phenomenon(a):
  enemy of science, 29
  regularities, 30
Phenomentalism, 58n35
Philosophers:
  favoring scientific notions, 87
  literary, 4
  literary criticism, x
  mystery in lives of, 91–92
  proper usage of terms, 76
Philosophical psychology:
  philosophy of mind (the term), 121
  terminology, 121
Philosophy:
  antiquity and worth, 4
  Carnap on logical analysis, 79
  Garrigou-Lagrange, 94
  literary expression, 2

organization of psychology, 121
origin; common sense, 7
psychology's association with, 118
reduced to interpretation, x
science and, *see* Science and philosophy
scientistic concept of, 4
teaching moral psychology, 127–28n3
teaching of positive psychology, 127n2
two conceptions of, ix
within science, 44–49
Philosophy of man, xv
Philosophy of nature:
  Aristotelian, 16
  becoming and plurality as problems, 16
  bibliography, viii
  definition, 56n27
  determinism, 38
  dissociation from philosophy of nature, 93
  division into treatises, 98n5
  First Cause, 15
  Frank and, 87
  genus and species, 99n5
  metaphysics, xiv
  mobile being the proper object of, 39
  object is nature itself, xi, 38
  order of abstraction, 98n5
  physical necessity, 43
  Plato, 58–59n39
  positive science abandons, 47
  psychology; positive science, 118
  Thomas Aquinas on, 129n5
  Thomism, xiv
  two states, 48
  unity of, 99n5
Philosophy of science:
  bibliography, viii
  leading idea of, 30
  location, xi
  study of soul, 129n5
  Thomism, 3
Photons, 42
Physical knowledge:
  abstraction, 103
  being capable of changing, 104
  definitions, xv
  division, 93, 95

meaning: whole of the sciences, 100
   object of, 100
Physicists:
   clarity and confusion, 1
   classical, 31
   cosmic image, 49, 65n62
   determinist vs. indeterminist, 46
   philosophize when crisis occurs, 48
Physics, 71, 79, 97n4, 118
   anti-ontological, 39
   classical, 86
   concept of efficient cause, 24
   crisis of indeterminism, 41
   Hahn on statements in, 79
   idea of determinism in, 40
   indeterminacy in, 38, 87
   mathematized, 24
   mechanistic, 86
   new concepts (contemporary), 41
   Newtonian formulas, 86
   object: mobile being, 104
   old and new, 86
Piaget, Jean, xvi
Piéron, Henri, 50n7
Planck, Max, xiii, 42, 44–47, 86
Planet, 83
Plato:
   philosophy of nature; myths, 59n39
   scientific object, 29
Playwrights, 116
Pluralism: epistemological, 90–99
Plurality, 11, 13
Poetry: lyric, 78, 79
Poets, 116
Poincaré, Henri, xii, 28, 30, 63n54, 67, 71n1, 85
Poirier, René, xix$n$20
Popper, Karl, viii
Positive psychology:
   applied psychology confused with, 129n6
   case: night vision, 123n1
   epistemological) aspects, 120, 121
   philosophical psychology and, 126n2
   possibility of, 120
   teaching of, 126–27n2
   tendency toward separation, 127n2
   term; only correct expression, 121
   two tendencies in, 126n2
Positive science(-s)(-ists), 2, 118, 120
   antinomic tendency, 48
   cosmic image, 48–49
   definition, 56n27
   detaching from philosophy, 128n4
   explanatory significance; proper cause, 21
   meaning of in/determinist ideas, 40
   measurement, 45
   philosophy of nature, 39, 47, 48
   prestige by association with, 127n2
   proper use of terms, 76
   rational appetite for being, 119
   relations: observation and theory, 79
   revolutions in, 49
   scientistic philosophy, 4
   technology, 122
   true knowledge from, xiv
   univocal science of all certain knowledge, 90–91
   use of terms, xi
Positivism:
   Carnap on, 78
   tendencies in, 57n35
Possibility(ies):
   eternal necessities, 30
   necessary laws of, 109
   possible thing; necessity, 26
Potency:
   Aristotle on, 53n19
   concept of, 54n19
   intelligible only in relation to act, 54n19
   origin of being, 17
   passive; material cause, 17
   preformation, 17
   radical misunderstanding, 54n19
   see also Being; Reality
Pragmatism, 125
Prayer: for weather, healing, etc., 63n54
Predicate:
   absolute and relative, 52–53n16
   distinction, 82
   sempiternal things, 57n33
   universal, 57n33
Predictability:
   meaning and, 10
   meaning (broad sense), 51n8
Prediction:
   causes placed in category of chance, 70

certain and exact, 44, 46
chance events, 44
Comte on, 67
concrete; pure science, 68
conditional, 69
cosmic image, 49
denial of successful, 60n48
essential function of science, 67
exact; determinism, 40
free reactions, 122
future contingents, 61–62n54
in science, xiii
man of action, 113, 116
natural events, 62n54
position of particle, 41–42
research on nature, 67
science dedicated to, 21
scientific, 44, 69, 71
system of initial data, 44
systematic knowledge, 69, 70
theoretical formulation, 68
three modes of, 69
Pride, 99n5
Prime matter, 104, 107
Probability:
  a priori and *a posteriori*, 37, 38
  definition, 33
  mathematical, 33–35, 37
  number of alternatives, 60n48
  physical, 33
  physical equi-probability, 34–35, 37–38
  Sorel's analysis, 71
  uncertain events, 35
Processes: isolated state, 68, 69
Prognostications, 85
Propaganda: free will and, 124
Propositions:
  conformity with reality, 79
  facts of experience, 80
  meaning, 77
  measurement; truth, 85
  metaphysical; meaning lacking, 78
  proof; rational evidence, 77
  scientific, 26, 47
  truth, 84–85
  verification; two kinds of, 77
Providence:
  chance and, 53n16
  explaining events, 15
  First Cause infallibly effective, 15
  fortuitous remains, 16
  fortuitous/chance events and, 52n16
  large variety of causes, 52n16
Prudence, 10, 100
Psychological testing: applied/moral psychology; cases, 123–24
Psychology:
  applied, 122–26
  applied and moral, xvi–xvii
  atmosphere of study courses, 111
  confusion with positive psychology, 129n6
  confused association with philosophy, 118
  disorder in study of, 111
  empirical science, 79
  epistemological situation, xv, xvi, 111, 125–26
  ethical neutrality, xv
  facts; common sense, 118–19
  facts; ontological interpretation, 119n3
  forms of knowledge confused under name of, 111–12
  general (the term), 122
  hiring drivers (case), 123
  ideas; representations, 21–22
  identity crisis, xvi
  knowledge of the singular, 116
  knowledge of the soul, 118
  language problem, 111
  man of action, 113–14, 116
  metaphysical self and non-self, 111
  metaphysical (the term), 121
  observation in, 120, 121
  philosophical, 126
  Piaget on philosophical, xvi
  positive concepts in, 119
  positive psychology exists, 118
  positive science, 118
  practical, 114
  practical (the term), 121
  proliferation of theories, 111
  psychic constraint, 124
  rational (the term), 121
  reorganization of teaching of, 126
  school programs, 126
  science of nature, xiv
  scientific (the term), 122
  speculative (the term), 121
  study of the idea, 39
  system-building, 118

terminology confused and changing, 118
theoretical, 117, 121
Thomism, 93
two practical psychologies: moral and applied, 122
use of the term in the singular, 126
vocabulary of, 121
the word; usage, 112
see also Soul
Psychotherapy, 123
Pythagorean relation, 13

Quality(ies):
  Aristotelian definition, 107
  definition, 107
  makeup of a whole, 107
  mobile things, 109
  proceeds from form, 108
  substantial form, 107
Quantity:
  Aristotelian definition, 107
  matter and form, 107
Quiddities, 96, 97n4, 109
Quine, Willard Van Orman, x

Rabeau, Gaston, 3, 65n62
Radiation: corpuscular nature of, 42
Rastignac, 116
Rational science, 67
Rationalism:
  certainty, 79
  chance and, 13
  Greek, 67
  realist; idealist, 72n4
Rawls, John, xvii
Reactions: determined, 122
Realism:
  Carnap on, 78
  naïve, xii
Reality:
  Carnap on problems of, 78
  concrete; composition, 67
  ideas and, 39–40
  logical propositions and, 79
  Marxist materialism, 72n4
  materialistic metaphysics, 13
  mathematical reading of physical, 86
  science of the really real, 59n39
  subject and predicate, 82
  two paths to knowledge of physical reality, x
Reason:
  being(s) of, 12–13
  practical reasoning, xvii
  rationalism and science, 72n4
Reductionism, xiv
Regularities, 30–31
Religion, 90–91
Renouvier, Charles, 118
Research, 67, 120
Resemblance: equation, 24
Responsibility: case; moral psychology, 123
Rorty, Richard, x

Sartre, Jean-Paul, xvi
Schlick, Moritz, 77
Schrödinger, Erwin, xiii, 31, 32, 33
Science(s):
  applied science, 68
  certainties, 25
  contingency-less world, 68
  co-ordination of ideas, 97n4
  critique of understanding, 48
  dealing with the universal, 100–101
  discovery; order, 30
  diverse natures fall under same science, 96–97n4
  division into treatises; unity, 98n5
  epistemology, 95
  evolution (three centuries), 92–93
  explanation and prediction in, xiii
  explanation of phenomena, xixn19
  explanatory necessity, n101
  formal and specific object, 97n4
  goal of, xii
  Greek rationalism, 29
  ideal; the real world, 24–38
  infused, in Christ, 97n4
  integration of science in law, 29
  knowledge of existential states, 25
  knows how of phenomena, but not why, 23
  lawful processes, 68
  mathematico-physical, 87
  mataphysical hypothesis as fundamental in, 45, 46
  microscopic experience, 27
  mixed sciences (theory), 97n5
  multitude of sciences; of scientific objects, 92

INDEX

nominal definition, 100
object of, 100, 104
ontological purification, 48
origin; common sense, 8
overestimation of, 91
phenomenon as enemy of, 29
philosophic foundations, 49
philosophic interpretation of data, 2, 28
physical and historical, 85
physical; Frank on, 86
picture of human reason (Goblot), 71n4
pluralist system; Thomism, 97n5
prediction and, *see* Prediction
proper, for each individual thing, 97n4
relation fo being of things, 47
seeks to know thing in itself, 45, 46n2
specific distinction of, 97n4
statistical/causal formulas, 31
systematic knowledge, 67–73
underlying forms of reason, 72n4
wisdom and, xvi
Science and philosophy:
    assimilation of philosophy by science, 93
    collaboration, 2
    common search for explanations, x
    distinct; akin; not to be separated, x
    distinguishing the "invisible," 84
    explanation and interpretation, x
    identical propositions; true/false, 95
    illusions (Frank), 86
    judgmental philosophers, ix
    kinship, ix
    metaphysics, ix
    necessity and contingency, 87
    philosophy as deriving from science, x
    philosophy within science, 44–49
    positive psychology, 118
    scientific character of philosophy, 91
    scientism, 4
    Simon's middle ground, x
    superior view of the philosophical, 95
Scientism, 78, 90
    decline, 4

Vienna Circle, 91
Scientists:
    metaphysical curiosity, 51n15
    philosophy of, 48
    raising philosophical issues, x
Self: science and, 128n4
Self-identity, 101, 108
Sensation:
    common sense, 105
    external; pure, 106
    mutability of the sensory world, 107, 108
    natural determinism, 115
    sensory powers, 106
    world of sensible qualities, 106
Sense perception:
    grasp of accidents, 59n39
    principle of causality and, 46
Sensible:
    proper; common, 106
    qualities, 108, 109
Shakespeare, William, 112, 128n2
Sign(s):
    definition, 74
    instrumental, 74–75
    intentional signs, 75
    object and the knowing subject, 76
    object made present to the mind, 76
    objectivity, 76
    observable aspects of things, 105
    signification: Vienna Circle, 74
    usage of the term, 74
    see also Meaning
Simon, Yves R.:
    on sensation, xv
    study of science and medicine, ix
Sincerity: moral psychology, 123
Skinner, B. F., xvi
Social engineering, xv
Social ethics, 66n62
Social philosophy, xv
Social sciences, xiv–xv, 66n62
Socrates, 92
Solitude, 112
Sorel, Georges, 57n32, 67, 68, 70–71, 96n3
Soul:
    animating role of, 121
    definition of, 129n5
    knowledge of the, xv–xvi

science of, important for morality, xvi
sciences of, 112
study of rational part of, 129n5
two causal systems in, 122
two theoretical sciences of, 117–18
see also Psychology
Speaking: suppressing possibility of, 48
Species, 97n4, 99n5, 103–104
certain and definable, 110n4
individual existences in, 26
plurality of, 27
Statement:
truth of, 85
words "able to be constituted," 83, 84
Statistics:
laws of nature, 32–33
statistical laws, 28, 31
Stoics: on causality, 12
Strauss, Leo, xv
Subject: distinction, 82
Subjectivity, 48, 49
Substantial form: quality and, 107
Sufficient reason, principle of, 34
Suggestion: hypnotic; post-hypnotic, 124
Supreme Good, 59n39
Synthesis: definition, 114
Systematization: positive psychology, 119, 121

Tautology:
logical deduction, 81
object signified by subject/predicate, 81–82
precise formulation of notion of, 81
Technology:
extended to man; dissent, 125
obsessive ambition for knowledge, 122–23
positive sciences, 122
Tendency (concept), 57–58n35
Terms:
use by scientists and philosophers, xi
see also Word(s)
Theology, 59n39, 98n5
Thermodynamics, 58n38
Thesis and antithesis, ix

Things: absolute and relative, 52n16
Thinking, *see* Thought
Thomas Aquinas, Saint, 52n16, 71n4, 96–99n5, 101, 104, 118, 129n5
on Aristotle's work on interpretation, 12
on future contingents, 62n54
human acts, xvii
logic of Aristotle (opusculum), 57n33
speculative sciences, 92, 93
Thomism; Thomists, viii, xii, xiv, 93
philosophy of science, 3, 4
system of sciences, 97n5
Thought:
general laws of being; observation, 79, 82
harmony with nature (Hahn), 80
practical, 14
subsistent, 20
suppressing possibility of, 48
Time, 20
Totalitarian states, 125
Toulmin, Stephen, xiii
Trust, 113
Truth:
common sense, 8
contemplation of, 71n1
eternal, 26
Hahn on, 84–85
historical/scientific, 85
imagination, 65n62
inferred from middle term, 96n4
magnificence of, 94
multiple, 95
pragmatic conception of, xiv, 85
theoretical psychology, 117
Vienna Circle, 84
Type(s):
De Koninck's system, 110n4
formulation, 105
individual existence; universal type, 116
prime matter, 104
universal, 105, 116

Understanding, 16, 97n4
intersubjective, 1–2
metaphysics of, 48
relating to self, 77
Unitary science, 91
Unity:

INDEX

causal, 12
diversity and, 54n21
plurality of causes, 12, 13
world, 16
Universal(s):
  abstraction, 108
  incorruptible, 103
  natural science, 99n5
  necessary object of science, 100–101
  nominalistic theory, 26
  physical knowledge, 103
  sorting whole from parts, 110n3
  type; definition, 110n3
Universe:
  absolute unity, 13
  center; end of motion, 65n62
  movements of greatest bodies of, 49n1
  no absolute beginning, 17
  order in, 52n16
  phenomenal, 29
  physical causality, 25
  present state, 49n1
  rationality; reality, 11
  reduced to single cause, 26
  regularity in, 31
  strictly unified, 24–25
Unpredictability:
  chance; physical events, 7, 10

Value neutrality, xv
Value statement, 78
Vegetable life, 98n5
Velocity, 42
Verification:
  Carnap on, 77
  two kinds of propositions, 77
  Vienna Circle, ix
Vienna Circle, ix, xiii, xiv, 1, 39, 77, 78, 91
  founder: Hans Hahn, 79
  theory of signification, 74
Virtual distinction, theory of, 16
Virtue:
  effective promotion, 125
  moral psychology, xvii
  preaching, 125
Vouillemin, Général Charles Ernest, 4, 58n37, 64n60, 74
Vries, Hugo de, 66n62

Weber, Max, xv
Whitehead, Alfred North, xiii, 65n62, 72n5
Whole:
  object, see Object
  psychology of man of action, 113–14
Will:
  divine government, 20
  of God: things done necessarily/contingently, 52n.16
  voluntary movement, 55n22
  weakness of, 114
Wisdom:
  science and, xvi
  scientistic philosophy, 4
Witness: false testimony, 123
Words:
  criticism of oral sign, 76
  instrumental signs, 75
  legitimate, 83
  meaning; distinction, 4
  omniscience, 81
  scientific statements, 83, 94
  teaching psychology, 126
  term(s) endowed with meaning, 76
  verifying the meaning of, 76
Work: idea or pattern in, 19
World:
  cause; sense-perception, 46
  denial of reality of (Carnap), 78
  diversity; unity, 16
  intelligibility, 16
  ontological meaning, 39
  principle causing action, 20

Zeller, Eduard, 129n5
Zoologists, 103